T0073607

FIRST DAWN

FIRST DAWN

FROM THE BIG BANG TO OUR FUTURE IN SPACE

ROBERTO BATTISTON

TRANSLATED BY BONNIE MCCLELLAN-BROUSSARD

THE MIT PRESS   CAMBRIDGE, MASSACHUSETTS   LONDON, ENGLAND

This translation © 2022 Massachusetts Institute of Technology

All rights reserved. No part of this book may be reproduced in any form by any electronic or mechanical means (including photocopying, recording, or information storage and retrieval) without permission in writing from the publisher.

Originally published as *La prima alba del cosmo*. © 2019 Mondadori Libri S.p.A., Milan

The translation of this work has been funded by SEPS
Segretariato Europeo per le Pubblicazioni Scientifiche

S E P S
SEGRETARIATO EUROPEO PER LE PUBBLICAZIONI SCIENTIFICHE

Via Val d'Aposa 7, 40123 Bologna, Italy
seps@seps.it—www.seps.it

The MIT Press would like to thank the anonymous peer reviewers who provided comments on drafts of this book. The generous work of academic experts is essential for establishing the authority and quality of our publications. We acknowledge with gratitude the contributions of these otherwise uncredited readers.

This book was set in Stone Serif and Avenir by Jen Jackowitz. Printed and bound in the United States of America.

Library of Congress Cataloging-in-Publication Data

Names: Battiston, Roberto, author. | McClellan-Broussard, Bonnie, translator. | Bartusiak, Marcia, 1950– writer of foreword.
Title: First dawn : from the big bang to our future in space / Roberto Battiston ; translated by Bonnie McClellan-Broussard ; foreword by Marcia Bartusiak.
Other titles: Prima alba del cosmo. English
Description: Cambridge, Massachusetts : The MIT Press, [2022] | Translation of: La prima alba del cosmo, © 2019 Mondadori Libri S.p.A., Milano.
Identifiers: LCCN 2021051209 | ISBN 9780262047210 (hardcover)
Subjects: LCSH: Astrophysics. | Cosmology. | Outer space.
Classification: LCC QB461.3 .B38 2022 | DDC 523.01—dc23/eng20220228
LC record available at https://lccn.loc.gov/2021051209

10 9 8 7 6 5 4 3 2 1

What really interests me is whether God had any choice in the creation of the world.

—Albert Einstein

CONTENTS

FOREWORD

This is the story about the greatest decades of advance in the history of astronomy, set amid a burst of innovation in both theory and technology. Your guide in this journey, physicist Roberto Battiston, shows us it was a time when the very definition of the universe underwent its most radical alteration. Few saw it coming.

At the start of the twentieth century, astronomers generally believed that the Milky Way and the universe were one and the same. The cosmos was a single, spiraling disk of stars set amid a void of unknown dimensions. There had been previous speculation that other galaxies existed beyond the borders of the Milky Way. In the eighteenth century, the noted British astronomer William Herschel viewed dozens of faint nebulae distributed over the celestial sphere and, like the philosopher Immanuel Kant earlier, briefly thought of them as island universes, separate congregations of stars. But this idea waned, taken over by the popular notion that many of these nebulae, especially those with a spiraling structure, were more likely baby solar systems in the making. But with a decisive observation in the fall of 1923, using what was then the largest telescope in the world at the Mount Wilson observatory in California, Edwin Hubble settled the debate once and for all. Finally able to measure the distance to a special variable star within the Andromeda nebula, Hubble determined that the nebula was indeed far-removed from the Milky Way. We were just one of billions of other galaxies spread over the depths of space.

This new, mind-altering discovery changed the thrust of astronomy completely. Once confined to planetary studies, solar research, and stellar classifications, astronomy now oversaw a far larger and more diverse cosmological vista. And with perfect timing, new ideas in theoretical physics arrived to provide needed insights into the universe's behavior. Astronomers had already noticed that most of the distant nebulae were fleeing away from the Milky Way, and Einstein's newly crafted general theory of relativity (which totally redefined our notions of space and time) provided an explanation. Applying Einstein's equations, the Belgian physicist Georges Lemaître determined in 1927 that space-time itself was expanding, which is why the galaxies were moving outward. They were going along for the ride. Two years later, Hubble provided the full observational proof.

Astronomers could even begin to contemplate the universe's very origin. By mentally putting the universe's expansion into reverse, picturing the galaxies drawing closer and closer together, they could imagine the universe emerging from a brilliant fireball—what came to be called the Big Bang. Particle physicists went on to explain how all the elementary particles and forces emerged within this cauldron of fire within the first few seconds.

New physics was also introducing astronomers to types of stars never before imagined. Using both Einstein's equations and the laws of quantum mechanics, physicists were forced to conclude that the heavier stars could shrivel up to the size of a city at the end of their life (neutron stars) or collapse altogether (black holes). At the same time, nuclear physics at last solved the longstanding mystery of how stars shine.

Over the succeeding decades, further revelations arrived once astronomers expanded their eyesight into other regions of the electromagnetic spectrum. Radio telescopes were erected around the world and found the first hints of quasars, young galaxies spewing the energy of trillions of suns due to the action of a supermassive black hole spinning wildly in each galaxy's center. Within the Milky Way, radio astronomers discovered pulsars, tiny neutron stars emitting periodic radio beeps as they whirled around. And permeating all of space was the leftover microwave radiation from the Big Bang, proof that the fiery explosion did occur nearly 14 billion years ago.

With the arrival of the space age, optical telescopes, X-ray satellites, gamma-ray telescopes, and infrared and ultraviolet observatories began to orbit the Earth and saw that the universe, long thought to be fairly serene, was actually suffused with titanic energies and explosive behaviors. X-ray sources provided the first indirect evidence that black holes existed. Perhaps the biggest surprise was the discovery that the universe was filled with some kind of added substance—a yet-to-be-identified dark matter—and a dark energy that is accelerating the universe's expansion outward. Ordinary matter—the stuff that makes up stars, nebulae, galaxies, and us—is less than 5 percent of the universe's contents.

Simultaneously, astronomers realized that collecting electromagnetic radiation was not the only means of studying the universe. As Battiston points out in this book, "The universe can be considered an immense laboratory in which matter and energy are continually subjected to the broadest range of experiments." Sensitive instruments embedded underground and beneath the Antarctic ice capture ghostlike neutrino particles emanating from both the Sun and the most violent events in the universe. Coupled with experiments conducted at such high-energy particle accelerators as CERN in Switzerland, cosmic rays arriving from deep space are helping particle physicists resolve nagging mysteries in their standard model of the components of matter. Battiston, former head of the Italian Space Agency, himself spent years developing special detectors to search for antimatter particles in space. These instruments operated aboard both a space shuttle and the international space station. He and his colleagues were trying to answer an outstanding question: why is the universe primarily composed of matter? How did the antimatter, which was inevitably created in the Big Bang, disappear? Or is there some of this bizarre material still out there?

Space-time itself is providing a new observational platform for astronomy. Starting in 2015, detectors erected in both Louisiana and Washington State (the two form the Laser Interferometer Gravitational-Wave Observatory, or LIGO) have been on the lookout for gravitational waves emanating from the spectacular collision of black holes and neutron stars. The VIRGO observatory, constructed by a European consortium, later joined the search from a location near Pisa, Italy. Gravitational waves are literally quakes in space-time's very framework. The ripples alternately stretch and squeeze space as they pass through at the speed of light. Right at the site of a clash, a six-foot man would be stretched to twelve feet, then within a millisecond compressed to three feet, before stretching out once again. Fortunately, after a journey of up to millions or billions of light-years to arrive at Earth, these waves are reduced to a quantum quiver, jiggling space by less than the width of a proton particle.

The fact that LIGO and VIRGO can even detect such tiny waves was a triumph of scientific ingenuity, which has led to an entirely new means of observing the universe. As the waves can pass through matter as if it weren't there, they offer astronomers the opportunity to peer into the very heart of an exploding star, to study the structure of a neutron star, and hopefully in the future to view the very moment of creation, by examining the gravitational waves released from the Big Bang itself.

Even as astronomers peered outward toward the edge of time, our local neighborhood was never forgotten. After decades of trying, astronomers in the 1990s were at last able to discover planets circling other stars within our galaxy. At first the exoplanets sighted were few and fairly large, the size of Jupiter or more. But with improved techniques and space-based missions, thousands more were revealed, and a sizable number are fairly Earth-like, with atmospheres possibly conducive to life. Battiston takes note that these discoveries have rejuvenated that age-old question: "Are we alone in the universe?" Could Earth have been seeded by space dust that contained hibernating life from another star system?

In the closing chapters, Battiston focuses on our new era of space exploration, where private enterprises are either joining hands with governmental agencies or in competition. Plans for getting back to the Moon, going to Mars, and miniaturizing Earth-orbiting satellites to monitor the globe are in the works. But Battiston, filled with an optimistic spirit, imagines an even more ambitious future: human intelligence forging an alliance with artificial intelligence to take us from our local stellar neighborhood into the universe at large.

Marcia Bartusiak

PREFACE

The three scientific revolutions of the twentieth century—special relativity, general relativity, and quantum mechanics—have had a profound impact on our view of the physical universe.

These powerful theories, together with increasingly accurate observations and measurements, have allowed us to understand and describe phenomena on energy and time scales incommensurate with our human ones. A new image of the cosmos has gradually developed, forcefully garnering our attention and becoming a profound part of contemporary culture.

During this time, the horizon of knowledge has extended tremendously in the direction of the infinitely small as well as toward the infinitely large. The scientific method has allowed our fragile species—in a random instant of the universe's evolution and from the arbitrary viewpoint of one of the billions and billions of planets in one of the billions and billions of galaxies of which it is composed—to impel our gaze toward the boundaries of knowledge, developing a narrative that has never in the course of human history been so accurate and, above all, shared.

In this book I want to present that story, accompanying the reader on a journey toward the boundaries of our knowledge of the physical universe and trying to sketch out what could lie beyond the limits we've explored so far. These boundaries involve time, space, and the properties of matter and energy, as well as the emergence of complexity on an ever-increasing scale. Boundaries to be approached with the boundless, childlike enthusiasm that animates scientists when they understand an essential aspect of a phenomenon but, at the same time, with the prudence of those who know that translating scientific language into a more accessible language inevitably results in the simplification of concepts. Above all, these boundaries should be approached with the humility of those who understand that what we know is only a minuscule fraction of what there is to know.

However, scientific progress over the last century has been so great that it's worth taking a few risks just to get a glimpse of the treasures that have been discovered and to try to imagine what still awaits us.

Personally, I have always been fascinated by borders, intrigued by edges, attracted by the discontinuity that exists between the frontier and the abyss, between the new and the old, between knowledge and ignorance; this is why I chose to become a scientist, and I have never regretted it. This curious wandering in search of an explanation for things is an essential aspect of science, an aspect that I wanted to render, involving the reader in a journey through dozens of orders of magnitude and in all directions.

Curiosity is characteristic of our species, a quality we all share to some degree. Continuity bores us, repetition sends us to sleep, the horizon excites us more than the sea. For scientists, this curiosity grows and expands into a passion, sometimes an obsession, and even a profession. Science aims to overcome the limits of our current knowledge wherever they are. Like explorers seeking new lands, researchers extend their exploration, searching out the best, most solid foundations upon which to build the most extensive and reliable vision of the world.

Science has developed due to the contribution of individuals or groups of researchers, sometimes quite large. This development has almost never proceeded in a linear fashion. Although its goal is to identify objective laws, the story of research is a story of people, of trial and error, of coincidence and of the unexpected. It is a very human endeavor, whether individual or collective, but the aim is to find results to which we assign a universal value.

In the forty years of my scientific career, I've had the opportunity to get involved with various areas of research, to participate in projects that seek to move beyond the frontiers of knowledge, and to meet people who have made or are making important contributions to the construction of our current and future vision of the world. In this book, I've chosen to present some of my own personal experiences as a researcher because I think they can help readers to understand the spirit in which researchers, now and in the past, approach the frontiers of knowledge. The closer we get to the present, the less certain we are of the roads we are taking. Researchers are a bit like explorers, setting off in search of one place and landing instead in a new world. But there can be no research without the spirit of the explorer, without the ability to take on the burden of the risk of failure, without the tenacity of the desire to follow lonely roads, perhaps for years, and without the human element of science—in all its grandeur as well as in its fragility.

It is only our very human qualities of tension and curiosity that drive us to try to understand what we see, which will allow the shadows we intuit, in the dawn that illuminates the frontier of new knowledge, to become clear and distinct images—our inheritance and our legacy.

Trento, June 25, 2021

1

LIGHTS BEYOND THE HORIZON

It's dawn. It is the magical moment lauded by poets, the intense moment of separation between a before and an after, between dark and light. The blooming of the first pallid light that anticipates the Sun's arrival. This is how dawn has been celebrated and considered for ages. The diffuse light that precedes the moment when our part of the Earth turns to face the Sun. Whether it's the first dawn that impressed us as young people or one of many, whether it's observed from the edge of the sea or from the port-hole of the International Space Station (ISS)—where it happens eighteen times per day due to the orbital velocity—or even imagined on the Little Prince's tiny planet, where sunrises and sunsets can be had at will, the moment of dawn subtly combines stillness and movement. It resembles the swing that has reached the apex of its arc, the roller-coaster car when, imperceptibly, it nears its first, precipitous descent to the unpredictable path that awaits it.

And yet we know that things are different than they appear. The dawn seems as if it were running to meet us, but we're the ones who are trans-ported toward the light, carried by the planet's rotation. In reality, we're the ones who are appearing on the horizon, not the Sun. We're the ones who finally edge our way into the luminous cone of our star, emerging from the shadow the Earth casts on itself every night. Similarly, science flips our perspectives, transports us toward an understanding that often moves us away from common sense.

If dawn is a beginning, this book would like to accompany the reader on a journey to discover a series of beginnings, revelations, changes in perspective. A journey toward understanding how our comprehension of things has changed, how we manage to see them from another point of view.

Science aims to push the limits of the known, wherever these limits lie. As the explorer seeks new lands, so research extends over time, as well as space, with the goal of understanding how things are made, what laws describe them, which cause produces which effect. It is precisely thanks to science that, in recent centuries, our view of the world and our place in the universe has changed in a radical way, offering us unexpected horizons. It took a colossal effort, the ingenuity and dedication of extraordinary individuals, together with the intense and determined work of a vast scientific community. However, the impressive result of this effort has turned out to be of inestimable value: we were able, step by step, after countless attempts and errors, to emancipate ourselves from visions and concepts—sometimes based on myths and superstitions, sometimes on deceptive and intuitive appearances—often passed down for millennia. Dawns of knowledge were gained at a high price, eclipsing and leaving behind movements in human thought that, in the light of new discoveries, proved contradictory or were surpassed by a refinement of rational thought or experimental verification.

This book accompanies the reader on a journey toward the boundaries of knowledge and even a bit beyond: we will explore both the largest and smallest dimensions of the universe, we will approach the origin of time, and we will take a look at the most distant future. We will ask ourselves if, by chance, life comes from very far away and whether it can continue its voyage thanks to future space exploration.

We will also try to understand which technologies have helped us arrive at some stages in the journey that has led us to better know and understand the cosmos. We will discuss some of the laboratory research that explores the laws of the universe.

The universe doesn't reveal its secrets easily. Quite the opposite. It requires those seeking answers to questions about nature to follow grueling paths. As if that were not enough, in the course of this exploration, rife with obstacles, we are often our own worst enemy. Ours is an extraordinary species, characterized by an insatiable curiosity and a singular capacity for discovery, both in the physical world and in the world of ideas. At the same time, we are deeply conservative, ready to fight tooth and nail to maintain every kind of cultural status quo, and capable of quickly interpreting any level spot on the steep path of knowledge as a unique and definitive end point. One of scientific progress's fundamental turning points was the Copernican revolution, which removed Earth from the center of the universe, where the ancients had placed it—despite contrary opinions emerging since Aristotle's time—and began a process of relocating it in the cosmos that still continues today. We lulled ourselves with this illusion for millennia, convinced that we were at the center of the world, perhaps created specifically for us, unique and unrepeatable, with the Sun and the stars revolving around us, bowing to such a marvelous, albeit imperfect, creature.

Just think, as early as the third century BCE, the astronomer Aristarchus of Samos had already proposed the idea that the Earth revolves around the Sun. He wasn't able to prove his thesis—which for this very reason remained neglected for twenty centuries—and this helps us to understand the extent to which our presumed centrality in the cosmos has constituted one of the most difficult barriers to knowledge to break down, a true sign of the passage from ancient to modern. A few years after the discovery of the Americas, which in its own way undermined the Ptolemaic model, the Earth remained tenaciously at the center: the planets, the Sun, and the sphere of fixed stars rotating around it. It took the 1609 observations of the great Galileo—confirming the Copernican hypothesis, a real bridge between Kepler's laws and Newton's *Principia*—to begin the progressive demolition of the artificial pedestal on which we had mistakenly settled. But even the grand Copernican revolution was still rooted in the prejudices of the time, in particular, a deep and widespread conviction regarding the immutability of the world, both terrestrial and extraterrestrial. In short, we were still dealing with visions deeply steeped in anthropocentrism, which saw the world around us as an end point and a stable condition. It took a series of sensational and innovative scientific achievements, which led to new breakthroughs in knowledge, to free us from these conventional beliefs. Starting with Darwin's theory of evolution in the second half of the nineteenth century, through Wegener's theory of plate tectonics at the beginning of the twentieth century, continuing on to Hubble's cosmology in the first half of the last century, we began to understand the rhythm of processes, inherent in both animate and inanimate nature, which we had assumed were stationary simply because they were very slow when compared with our human scale.

Today we know that we live in a universe in the making, within which both living and inanimate elements have evolved and mutated relentlessly for billions of years. Its story is told in a book of which many chapters are still to be written, while the content of others, already written, we have not yet been able to fully decipher. A book that we can read, as Galileo said, if we know the alphabet of mathematical characters with which it is written. A narrative in which our story and that of planet Earth may take up only the space of a single, tiny paragraph—as much as it may seem crucial to our eyes.

The exploration of this universe is studded with attempts, tensions, expectations, and doubts but also with exhilarating successes, moments that have proved decisive in tearing away the veil of ignorance, and which have paved the way to a greater level of knowledge, with sometimes unpredictable consequences. On the other hand, scientific and technological revolutions in the fields of physics, biology, space exploration, the study of the complexity and exploitation of the power of modern information technology (IT) technologies have opened and are still opening new perspectives for us, at an ever-increasing pace. Every day we are receiving new information about important scientific achievements and advances, resulting from the efforts of hundreds of thousands of researchers working in every part of the world. Perhaps someday someone will highlight the limitations and shortcomings of the science being built today, but that is the basis of scientific progress. We are certainly experiencing an exciting phase in which we can intuit a multitude of knowledge horizons toward which we are moving rapidly.

If the dawn is the tension between the appearance of a new thing and that thing not yet being fully realized, it seems to me an excellent metaphor for research, which is a continual approach to new answers that lead to new questions: we work on a constantly renewed understanding of the universe that continually opens even broader scenarios of questions and problems.

2

TRAVELING COMPANIONS

As always happens when we set out on a new adventure, we have to pack our bags carefully, understand which things to take with us and who will be our traveling companions. According to many, the twentieth century should be remembered as the century of physics, which developed in an astounding way starting from the discoveries of Maxwell, Planck, Einstein, Bohr, Schrödinger, Heisenberg, Pauli, Dirac, Fermi, and Hubble, to name just a few of the giants who contributed to the progress achieved in the fifty years between the end of the nineteenth century and the beginning of the twentieth. By studying the atom, its nucleus, and the elementary particles, physics has been dedicated to the examination of the infinitely small, dealing with dimensions up to the order of a billionth of a billionth of a meter on an experimental level, discovering the unexpected limits but also the incredible possibilities of quantum mechanics. At the end of the twentieth century, this evolution in knowledge led to what is known as the Standard Model of fundamental interactions and elementary particles, a modern triumph of the Galilean method, culminating with the discovery of the Higgs boson at CERN, which we will discuss later.

Astrophysics and cosmology have also made enormous strides over the last hundred years. Ever more powerful instruments, which have been operating directly in space for just over fifty years, have allowed us to explore our universe with an extraordinary degree of detail and to extend our observations all the way back to the Big Bang, which is to say 13.8 billion years ago. Recent data from the Planck satellite, based on information obtained from the observation of a young universe—when it was *only* 379,000 years old—confirms the validity of the Cosmological Standard Model, the macrocosm analogue to that of the microcosm: this model is based on the idea of an exceptionally violent expansion, none other than the famous Big Bang, which we try to understand using the known properties of space and time, of matter and energy, in conditions dominated by quantum mechanics.

The infinitely small, which in practical terms means one billionth of a billionth of a meter, and the infinitely large, which means the distance traveled by light in a few tens of billions of years, are the experimental limits we face in physics and in astrophysics today. Even if we are dealing with more than forty orders of magnitude overall (multiply a number by ten and you will have an order of magnitude greater), between the extremes of the dimensions that we can study, we are still a long way from the infinite to which mathematics has accustomed us. The same is true for limits in the measurements of other physical quantities, such as time and energy. These are the limits of knowledge we are trying to overcome in order to see what happens beyond them, how nature behaves once we cross these frontiers, if the laws we have deduced on this side of the pillars of Hercules also apply beyond them.

Today, these limits seem truly difficult to overcome, but it must be said that we have already had such a sensation in the past, and we have been wrong. The distances that we are currently able to observe are almost certainly less, probably quite a bit less, than the size of the universe. In the time that has passed from the Big Bang to the present day, light has been able to travel only a limited distance, beyond which information from other areas cannot yet have reached us. There is therefore no reason to think that the universe does not extend further, exponentially further, than we are able to observe.

A similar argument applies to those dimensions that are increasingly smaller. To explore them, we need increasingly powerful microscopes as we slowly reduce the scale of what we intend to study. Modern microscopes are called particle accelerators, like those used at CERN for the discovery of the Higgs boson. But, as powerful as they are, their effectiveness is limited by the energy available in the particle beams used in these machines. In some cases, an attempt is made to overcome these limits by exploiting the very high-energy radiation coming from the depths of the cosmos. Where experimental analysis still cannot reach, theoretical study proceeds, which aims to understand if space-time is discrete or continuous, or how we can deal with the quantum fluctuations that dominate the smaller dimensional scales.

But we have come a long way! Only a hundred years ago we did not know of the existence of other galaxies; in 1800 we did not think that there could be stars more than 13,000 light-years away. In 1500 we were still convinced that the Sun revolved around the Earth.

Everything's all right, then? Not exactly. Despite these enormous advances, in a certain sense we are still not very far from where we started. The questions we ask ourselves are very similar to those asked by the philosophers of ancient Greece. What is matter made of? Is it infinitely divisible or not? Is the world as a whole infinite? Do things always remain cyclically the same or does the universe change? Do we understand reality by observing it or does thought have to grasp something that lies behind appearances? However, the context has changed, thanks to the development of experimental science. Four centuries after Galileo, today we know how to question nature and how to read the mathematical characters with which its book is written.

3

THE LAWS OF THE COSMOS

Today we can provide answers to the big questions, which we would have considered far-fetched or arbitrary yesterday, because they are based on a deeper knowledge of the nature of things and the laws that regulate the cosmos. Some general laws of physics are the solid foundations upon which the various levels of our knowledge about reality are based: special relativity, which extends Newtonian dynamics to conditions in which relative speeds are close to those of light; general relativity, which extends universal gravitation to the relativistic case; and quantum mechanics, which rigorously describes the characteristics and limits of the measurement process at the microscopic level. Then we have the laws of statistical mechanics, in particular, the second law of thermodynamics, which accurately describes the disorder of systems composed of many particles and its unstoppable growth with time. These fundamental laws provide for the existence of physical quantities that are conserved at the microscopic and macroscopic levels, dictating the rules that must be respected in every moment, place, and condition in the universe: energy, momentum or linear momentum (the product of mass times the speed of an object), angular momentum (the rotational equivalent of linear momentum), and electric charge. There are other quantities that are conserved, as far as we know, but which may not be in special conditions yet to be observed. For example, the number of particles called leptons, such as the electron and the neutrino, and their corresponding antiparticles, which have the opposite lepton number, are conserved, as are the number of the particles known as baryons, such as the proton and the neutron and their antiparticles, which have the opposite baryon number. There are, however, no conservation laws for other types of particles, such as those carrying elementary forces; this applies in particular to photons, the mediators of electromagnetic forces that can be created or destroyed without any counting problem.

This solid collection of physical laws has resulted from the work of generations of scientists. Today, we enjoy a perspective made possible by the fact that we are standing on the shoulders of giants; a point of view, it's important to remember, that we have gained over a very short time, the classic "blink of an eye," when compared with the billions of years the universe has existed.

Our minds have become accustomed to reasoning about time in billions of years, in units of light-years to measure length—a light-year is equal to about 300 times the diameter of the heliosphere—and today our language includes words and concepts that can cope with the very large, the very small, the very slow, and the very fast. Thanks to all of this, the history of our universe takes the form of a thrilling story, all the more fascinating because it is verifiable, in whole or in part, using instruments and methods that are the product of human intelligence, making it possible to share with humankind both today and tomorrow. It is a story that we will pass down to future generations along with the tools to continue with scientific investigation and to add new pages and chapters to nature's book.

In light of our current understanding of the laws of physics, the ingredients to describe the universe are relatively few: forces, energy, and mass. Time. Space.

With the theories of special and general relativity, on the one hand, Einstein has given us evidence of the equivalence of mass and energy; on the other, the deep connection between space, time, and gravity. In this way, the brief list of fundamental ingredients appears to be, from a conceptual point of view, further reduced. The fact that the list is so short does not mean that these concepts are simple or intuitive, as we will see later. Seen through the eyes of modern science, they reveal surprising characteristics, missed by past thinkers, that are, however, absolutely relevant to describing our universe.

Let's think about the concept of force. We have an intuitive experience of what a force is: something that forces an object to move in a different way than it would in the absence of that same force. Newton understood this perfectly well with his principles of mechanics in the seventeenth century. Since the era of the great English physicist, quantum mechanics has extended this idea, introducing the idea of force fields that fill the space between interacting charges, transmitted by suitable particles called *mediators*, which interact with matter. If we are thinking of an empty space and of long-range interactions, quantum theory gives us a profoundly different description of what a force is. We currently have evidence of a limited number of fundamental forces, four to be exact: gravitational, electromagnetic, weak, and strong. We do not know why there are four and why their intensity and range of action are enormously different. However, we do know that it is thanks to these forces that matter is organized in the universe at every scale.

As for mass, it has a central role in the modern description of physical reality. Since the time of the ancient Greeks we have been led to think of matter as formed by small, indivisible particles, shaped like a point and having weight. It took Newton to point out the difference between mass and weight. In his second law of dynamics, *inertial* mass is that property of a body that binds force to acceleration. In the case of universal gravitation, however, *gravitational* mass is the charge that generates and is subject to the force of gravity. In his general theory of relativity, Einstein unifies the two types of mass, attributing to it, through gravity, the fundamental role of warping space-time.

In developing the special theory of relativity, Einstein noticed another surprising property of mass, that it could be transformed into energy and vice versa. The famous equation $E = mc^2$ expresses precisely that: if we take a certain quantity of kilograms of mass and we multiply them by the speed of light squared (a very large number), we get an (enormous) amount of energy (which is measured in joules). Obviously, this transformation can take place only under special conditions, such as those that occur in nuclear fission or fusion reactions. The fact remains that one can look at the mass of particles as a container holding an immense amount of energy that, under the right circumstances, can be released. In addition, quantum mechanics treats elementary particles with mass as if they were waves, adopting a surprising probabilistic description that is not easy to fathom, but which works like a charm.

With regard to energy, we have an intuitive idea linked to the motion of matter (kinetic energy) and the propensity to do work but also to absorb it. Energy can be positive or negative. Just think of kinetic energy, positive and due to mass in motion, or of gravitational potential energy, negative in the case of two bodies linked to each other by gravity. We will see how essential this aspect proves to be in defining the first moments of the universe.

But there is no end to the surprises. The most profound conceptual revolution that we have witnessed is that having to do with space and time, unified in Einstein's theories of special relativity and general relativity. Time, as we know, has been dealt with by many philosophers and thinkers throughout history. From Saint Augustine, who stated, "What then is time? If no one asks me, I know; if I want to explain it . . . I do not know," to Henri Bergson, who separated conscious time, based on *duration*, from the series of juxtaposed *instants* characterizing external time. From here we go naturally to Immanuel Kant, who argued that time as well as space were not concepts derived from experience but rather conditions *a priori to experience* itself. Einstein's description of time, the only one we will be dealing with in this book, abandons all elements related to characteristics of human thought or states of consciousness. In an extremely pragmatic way, it brings the concept of time (and space) back to strictly defined procedures in the physical universe, based on rays of light, clocks, and rulers, aimed at establishing precisely what a temporal (or spatial) *interval* is. Einstein's approach leads to a geometric description of time and space for which the intervals are no longer absolute but relative. Measuring these intervals is similar to observing a geometric shape from different angles: in the case of time (and space) the angle of the point of view is replaced by the relative speed at which one observer is moving with respect to another. In the presence of a gravitational field, we factor in the effect of the curvature of the fabric of space-time induced by the presence of masses. In the appropriate conditions, such as the proximity of a black hole, the flow of time stops and therefore time may even *disappear*, an aspect that no one had ever considered before Einstein's theories.

And here we are, to top it all off, with space. The thing that appears to be the most evident and obvious is perhaps the most mysterious and difficult to comprehend. First, physical space is something that has properties, that is not simply a container of objects. Masses move and light is transmitted within space. The vacuum hosts the ferment of quantum mechanics' virtual states. Space can be folded on itself, closing itself in such a way that it excludes any need for external space from which to be observed. But space has many other properties, expressed in Einstein's general relativity equations. One among them is its *metric*, the quantity that determines the scale of distances. To understand what it is, let's imagine a sphere with a given radius: on the surface we put a placeholder every ten meters. Now, imagine that the radius of the sphere can be changed: if we double it, the distance between every pair of placeholders doubles; if we halve it, that distance is proportionally reduced. The distance between two placeholders gives me the metric of the sphere's surface. Changing the curvature does not change the number of placeholders, it only changes the distance between them. General relativity describes the effect of gravity as an influence on the metric of space-time; where gravity is very intense, the curvature radius is very small and space is very deformed, and vice versa. However, energy conservation is always at work. The accumulated energy in the curvature of the metric corresponds to the energy of the gravitational field; as the metric changes, there is an energy exchange with the masses present in space-time. What happens to the objects placed on the sphere's surface if the surface's curvature radius changes? Since these can be approximated as point-like objects compared with the sphere, which is much larger, they do not experience any internal deformations. It is only their distance from faraway objects that changes, and if the change in the curvature radius develops continuously in time, the further away two objects are on the sphere, the larger the relative speed between them.

Because the metric is a geometric feature of space, and not a physical quantity like mass or energy, it enjoys another absolutely unique property: the speed at which it can change is not subject to the limits of special relativity. Thanks to the metric and its properties, Einstein's space-time has gained a fundamental degree of freedom that wasn't present in the construction of Newton's dynamic.

Einstein's space is as deformable as a trampoline at the amusement park. Metric variations have accompanied the universe for the entire course of its existence, at strongly varying rates in the different phases of its evolution. As we will see, when the universe emerged, during the extremely short phase of what is known as inflation, the expansion of the metric resulted in an expansion in the volume of the universe at a rate much faster than the speed of light. In addition, after several billion years of expansion at a constant speed, about 5 billion years ago the metric's expansion rate resumed accelerating to cosmological distances, that is, comparable with the dimension of the current universe. Finally, it can be modulated, as in the case of gravitational waves, which are alterations in the metric induced by the violent acceleration of large masses. These waves were predicted by Einstein in his general theory of relativity in 1915 but were only directly detected a century later.

Therefore space-time, mass-energy, and fundamental forces are the main players in the grandiose spectacle of the universe, within the framework of the fundamental laws that regulate their interaction. These elements form the foundation of the enormous variety in the universe, in its multiplicity of forms, its complexity—which lies not in the repetition of many similar structures but rather in the infinite use of a few elements to create an innumerable quantity of different structures. And it is precisely in the context of complexity, the emergence of order from a universe apparently destined for increasing disorder, that we can introduce numerous themes science has dealt with in an original way that stimulates our curiosity, for example, the origin of life on our planet and the evolution of living beings, the presence of other forms of life in the universe, and the study of the laws governing the universe's complexity.

In the next chapters we will touch on some of these topics while making no pretense of being systematic but rather trying to intrigue the reader with respect to the overall picture. Today, scientific knowledge is increasing at such a rate that it becomes impossible to describe the continual progress, about which we are inundated with information on a daily basis, in a systematic way. In order to find our bearings in a world where science and technology play an increasingly relevant role, it is important to try to get a concise idea of what the key elements of this process are, separating the advancements in our understanding of the fundamental elements from the huge number of wondrous scientific and technical developments, which do excite our curiosity, but don't necessarily satisfy our desire to understand.

4

KNOWING THAT WE DON'T KNOW

Are there limits to knowledge? How much do we know about what we don't know? Of course, nobody knows the answer to this question, even if we have periodically succumbed to the illusion of having reached the final chapter in the book of nature. The scientific revolution that leads from Galileo to Newton, introducing the experimental method, on the one hand, and the principles of mechanics, on the other, has proved powerful and effective in dealing with the widest-ranging areas of physics and the most diverse dimensional scales. There began to be confirmation of the idea that an intelligence, which knew all of the forces in play and the respective positions of all of the parts that make up the world, would be able to predict the future with absolute certainty, deriving it from the past. In the seventeenth century, God was seen as the "watchmaker" of the universe, while science gave humankind the ability to reveal its mechanisms and control those parts of most interest to us, constituting what we would call a mechanistic vision of the universe. This concept still has a strong hold on the collective imagination, and not only in the field of exact sciences.

The advent of quantum mechanics, in the early decades of the twentieth century, completely changed our way of looking at things, creating a crisis for the mechanistic vision, at least as an exhaustive explanation. The study of microscopic matter—atoms and subatomic particles—forced scientists to a deeper analysis of the significance of the physical phenomena measuring process. This analysis led them to the astonishing discovery of the innermost probabilistic foundation of the physical world. Add to this the equally disconcerting fact that only observable quantities can be discussed within a physical theory. During the same period, Heisenberg set forth his famous uncertainty principle: in nature there are pairs of variables—for example, the position of a material object and its momentum (mass multiplied by speed)—that cannot be measured simultaneously with an arbitrarily small degree of accuracy: the measurement errors of the two quantities, multiplied together, are inevitably larger than a constant, Planck's constant. If we think of a bullet fired from a rifle, this principle is confirmed; if we try to precisely define where the bullet is at each moment in the course of its trajectory, we begin to lose track of the speed at which it is moving and vice versa. For a bullet the effect is negligible, but in the case of an electron this effect places an insurmountable limit on the definition of its trajectory between two points. For all intents and purposes, the electron acquires the properties of a wave.

On the whole, quantum mechanics is a solid theory; the very structure of atoms directly depends on the uncertainty principle. Nevertheless, Einstein never accepted this way of looking at things, making all kinds of objections over the course of more than thirty years, all regularly dismantled, sometimes with studies and research lasting decades and concluded after his death.

Quantum mechanics is a valuable tool for the exploration on which we are going to embark. Its foundations challenge mechanics at every turn and push us to reflect on what it really means to understand reality.

But there are also other factors that define what we know, compared with what is unknown. The boundaries of knowledge are constantly moving, but in a discontinuous, sometimes fragmented way: it is difficult to see a global strategy, an action aimed at unifying all knowledge. Some areas of knowledge can develop very quickly and then come to a halt for long periods. Decisive advances can take decades or even centuries. Today's research is no longer, as it once was, the preserve of a few, isolated scientists or circumscribed communities of scholars. Research is now a social activity involving hundreds and thousands of people around the world. However, even in areas where we are proceeding systematically and with enormous organizational and financial efforts, such as the physics of elementary particles and fundamental forces, we are currently in rough waters. The goal of unifying all kinds of elementary forces and forms of matter in a single overall framework characterized by compact and elegant mathematical laws and symmetries is ahead of us, but, for now, difficult to achieve. We would need indications and experimental proof to show us the way forward. Even at the level of cosmic observation, physicists and astrophysicists are grappling with a universe of which 95 percent appears to be dominated by a component of dark mass and dark energy about which we know very little, apart from the fact that it produces observable effects on the motion of stars and galaxies.

Reflecting on this situation, we cannot but think back to what Socrates said more than 2,400 years ago. He was hungry for knowledge but for this very reason he had the impression that knowledge was constantly slipping through his fingers. In Plato's early work, the *Apology of Socrates* (from which we learn much of what we know about his master), he tells us that Chaerephon asked the priestess of Apollo's oracle at Delphi who was the wisest man, to which she responded: "Socrates." Knowing this wasn't true, Socrates wanted to show that the oracle was wrong. So, he went to talk to those who had a reputation for wisdom: artisans, poets, politicians. But those with whom he spoke, put to the test by the philosopher's arguments, showed the limits and contradictions in the depth of their knowledge, revealing themselves as both ignorant and unaware of their own ignorance. Socrates then understood why he should be considered the wisest of men: he was the only one who knew that he didn't know.

A similar awareness of our own limitations pervades contemporary science. The idea of fallibility is intrinsic to modern scientific method, based on hypotheses and theories subjected to continuous experimental verification. A scientific theory, however impressive and important it may be, can be falsified by a single experiment; something that has happened to many scientists, including Einstein. What best suits the scientist is an attitude of conscious humility as they seek to progress in the research and theoretical interpretation of experimental data. This is especially true considering, in the history of science, how strong an influence our innate tendency to project our own preconceptions and categories onto nature has had.

We have noted how slowly the consciousness of our position in the cosmos developed. It is not enough to learn that we are on a minor planet, orbiting a small star, in an ordinary galaxy. Over the last fifty years we have realized that even the matter of which we are made—protons, neutrons, electrons, photons, and neutrinos—is less than 5 percent of the entire sum of what is present in the universe. The remaining 95 percent is a dark and mysterious ocean of matter and energy in which we naively navigate, but that shapes the visible forms of the entire universe. The progress of scientific thought has pushed us, it would seem, into a remote, dull corner of the cosmos. However, and it is a paradox, by definition we are at the center of an expanding sphere and the edges of it are cloaked with the radiation echo of the Big Bang. From our position, only seemingly "privileged," we admire a universe that, if we want to put it bluntly, has no center.

When I reflect on these aspects, I find it difficult to resist thinking about the meagerness of human activity, little more than cast-off dust, in the immensity of the cosmos. I think of this fragile living species, confined to a small planet, made up of individuals who are active and productive for a very short period of time, within an enormous universe that has been evolving for billions and billions of years. A species that, as Carl Sagan masterfully describes in his book *The Demon-Haunted World*, is dominated by passions, superstitions, and fears—the result of an ancient and persistent evolutionary and cultural heritage, in which the critical thought of Greek philosophers took shape just over 2,000 years ago and the scientific method began to influence the worldview only a few centuries ago. Against this background, the fact remains that today—despite our limitations—we are able to observe and study the cosmos, its origins, and its evolution. We are able to critically analyze and reason about the innermost structure of matter, forces, space, and time.

To the amazement of those who contemplate the beauty of everything that surrounds us, the scientist adds another layer of wonder in the very act of discovery and knowledge. We are astonished by the universe, but also by the explanatory effectiveness of a formula or a theory that we can use to describe it. It is as if we feel ourselves to be at the center of the universe not out of ignorant pride, but rather out of our awareness of how valuable it is that we can know it.

Special, precisely because we know we're not.

5

ELEPHANTS IN THE ROOM

We are about to embark on a journey into the universe, one that will lead us to explore some of the outer realms on the frontiers of knowledge. A ragged frontier, something like the tortuous path of China's Great Wall, up and down over ridges and mountains, dotted with forgotten outposts—rarely visited or jutting out over dizzying abysses—but no less necessary in defining the perimeter of knowledge. As we saw with Socrates, knowledge is all the more profound when it includes knowledge of our own limits. In addressing the question of what "substance" is in his *Essay Concerning Human Understanding*, John Locke tells of an Indian sage who, having stated that the world was supported by a great elephant, was then asked on what the elephant rested. He answered: a great tortoise. But when he was asked what supported the tortoise, he replied: something, he knew not what.

This famous philosophical anecdote is useful to us, by analogy, to reflect on how science, more frequently than one might imagine, is based on assumptions that are as fundamental as they are incomprehensible. A striking example is the theory of quantum electrodynamics (QED). It is a powerful formalism in which electromagnetism, special relativity, and quantum mechanics have been fused together to obtain a theory capable of predicting the phenomena involving electrically charged particles with extreme accuracy. QED theory was developed through the contribution of some of the last century's greatest theoretical physicists, especially Tomonaga, Schwinger, Dyson, and Feynman. It makes it possible for us to understand and predict phenomena that go from the behavior of electrons in atoms to the production of new elementary particles to particle accelerators such as the Large Hadron Collider (LHC) at CERN, which we will return to later. It is a very elegant theory, which deals with space and time, particles and fields, interactions and dynamics, while respecting the principles of quantum mechanics and special relativity. The predictive capacity of QED is exceptional. The detail of the interaction between electrons and photons can be calculated with an accuracy that, in some experimentally verified cases, has been better than parts per trillion, making it the most accurate physical theory in existence.

At the same time, it is surprising how this theory is plagued by such a number of divergences in calculations as to make it impossible to obtain a finite value for fundamental quantities, such as the mass or charge of the electron, without resorting to the elimination of quantities that are infinite in the calculation itself. This result is obtained by a procedure known as *renormalization*, through which the subtraction of two infinite quantities allows us to obtain a finite quantity that corresponds to the physical quantity in question—a horrible procedure for a theory capable of describing infinitesimal details with extraordinary precision. In 1975 Dirac stated that he was deeply dissatisfied with a theory in which rather than "neglecting a quantity when it turns out to be small—we reject it just because it is infinitely large and you do not want it!" In 1985 Feynman was still calling renormalization a "shell game" and suspected that it was "not mathematically legitimate."

Another example of the limit of our current knowledge is the question of vacuum energy. As we will discuss later in more detail, the quantum vacuum, far from being "empty," contains random fluctuations of particles and force fields that cause experimentally observable phenomena. It is precisely because of the vacuum's properties that, in a more or less long but calculable time, an excited atom reemits the absorbed energy and returns to the ground state. Like when, at the end of a run, your heart is beating faster and you are hot, but, after a bit, your body returns to its resting state. The presence of quantum fluctuations is equivalent to the presence of vacuum energy, which should also have a large-scale effect. As we will see later, this corresponds to a term, the cosmological constant, initially introduced by Einstein in his original general relativity equations to explain the existence of a stationary universe. Einstein later regretted introducing this term, declaring this decision his biggest blunder. The cosmological data provides a vacuum energy value of about one billionth of a joule per cubic meter (one joule is equal to about the work it takes to lift a book weighing 1 kilogram by 10 centimeters). However, the calculations of vacuum energy made using QED lead to a much, much larger result: a factor of 10 followed by 122 zeros larger. Physicists call it the most mistaken theoretical prediction ever made. It's another case of "infinity" in which we navigate without understanding its meaning. The infinities in QED or the inability to calculate the value of vacuum energy are examples of what we can call an "elephant in the room," a problem so relevant that it cannot be ignored.

Science has plenty of elephants roaming through its rooms. We shouldn't be surprised. From this point of view, scientific progress looks like an unending Sherlock Holmes investigation. In physics we have discovered that dark energy contributes about 73 percent to the mass/energy balance of the universe, even if we don't know what it is, and that about 22 percent of it is matter, also called "dark," about which we know even less. The matter and energy we know about are therefore a very small fraction of the whole, less than 5 percent. If we look at genetics, we see something similar. We know, for example, that about 80 percent of DNA has no protein-coding properties, which is why it is often referred to as "junk DNA." Despite several attempts, we have not yet been able to understand the evolutionary origin of this piece of the genetic code. In short, the part of DNA of which we know the properties represents only the tip of the DNA iceberg, the submerged part of which appears, nowadays, to be useless.

Then, when we think of a classic economy, which developed treating nonrenewable energy resources (coal, oil, natural gas)—built up on our planet over hundreds of millions of years—as raw materials, their limited supply has not been calculated into their value, over and above the cost of extraction and market availability, as well as not calculating the effects of their consumption, since the processes that produced them cannot be replicated or restored: a cultural misstep that has generated a series of consequences and had a negative impact on the environment. We started seriously and systematically studying the concrete cost and value of exploiting nonrenewable resources only a few decades ago. And that's not all. It has been calculated that the industrial economy absorbs almost twice as much value from the "invisible" reserves of our environmental and climatic ecosystem as it creates. What does that mean? Something very simple: that, today, we are consuming environmental resources at a rate that is 1.7 times faster than the planet is able to renew them.

Every time we encounter situations in which there is an elephant in the room, in which an overwhelmingly huge problem is more or less consciously ignored, it is reasonable to think that, sooner or later, there will be a day of reckoning. In a socioeconomic context it might be called climate change; in the scientific field it could be called progress or a revolution, depending on the effects that manifest. Einstein's greatness was that he proved to be particularly good at identifying and dealing with the elephants in the room of Newton's and Maxwell's physics, and at proposing radically innovative solutions that changed modern physics.

The existence of Brownian motion (that is to say, the disordered motion of small particles present in fluids or fluid suspensions, observable under the microscope, but also in suspended dust illuminated by a ray of light) was known since the time of Lucretius, who mentions it in *De rerum natura* (first century BCE). Einstein was the first to deduce, from the measurements of such an apparently trivial phenomenon, that matter is particulate, determining Avogadro's number (which corresponds to the number of atoms present in a given standard quantity of a substance, which corresponds to its atomic weight expressed in grams) and the size of atoms.

Physicists at the time realized that Maxwell's electromagnetism equations predicted an absolute speed, equivalent to that of light in a vacuum, but they could not understand how this could be in agreement with Galilean transformations between reference frames in relative motion, the cornerstone of Newton's mechanics. Einstein understood that it was necessary to change the Galilean transformations when the speeds involved approached the speed of light, making Newtonian space-time flexible.

The work for which Einstein received the Nobel Prize had to do with explaining the photoelectric effect, according to which it is possible to extract electrons from certain metals, properly illuminating them with a beam of photons. At the time, everyone thought that the phenomenon was linked to the intensity of the incident light. Only Einstein understood that it was linked to color and therefore to the energy of the light. This is how he introduced the concept of the photon, a particle able to carry a given quantity, a *quantum*, of energy—a powerful idea that made a decisive contribution to the birth of quantum mechanics.

Finally, Einstein understood that space and time were shaped by gravity, which distorts its structure according to precise mathematical laws. A body in free fall in a gravitational field loses all ability to experience the presence of gravity. The gravitational field *is* the space through which bodies move and *is* the time measured by clocks.

In short, returning to our original metaphor, encountering an elephant in the room is lucky: we must not stop thinking about it. It is possible, and probable, that it will be the starting point of the next scientific revolution—an opportunity not to be missed.

6

FIAT LUX?

Beginning, heaven, earth, the deep, wind, water, formless void, darkness. And then light. On the first day of the biblical creation narrative, the universe already contained all of the elements except light. The divine act, on the second day of creation, was in separating the light from the darkness. To be more accurate, we should actually say, in creating light, considering that darkness is only the effect of its absence. But what really happened? No one was there, so no one can speak from direct experience.

The archetypal idea of a beginning has always accompanied humanity, but the idea of a stationary universe has also been a constant influence on how we think, starting from the concept of an eternal cosmos so dear to Aristotle and enormously influential over the following centuries. Einstein, for one, liked it very much. For millennia, in an attempt to address the issue of origins and provide a narrative of the creation of the cosmos and its laws, a cosmogony, humanity has resorted to mythical tales, to analogies, to the combination of powerful elementary ideas that described the indescribable. We find traces of the idea of cosmogony as early as the fifth century BCE, in Leucippus, considered the father of atomism by Aristotle and to whom the *Megas diakosmos* (Great Cosmogony) is attributed, inspiring his student, Democritus, to write the *Micros diakosmos* (Small Cosmogony). However, different cosmogonies characterized by archaic myths and stories are legion, and date back to well before the Greek philosophers. From the ethno-anthropological point of view, it is useful to analyze the language and concepts used by civilizations distant from us in time and space, as they allow us to understand their theological, social, and cultural aspects. In the absence of scientific language and theories, from time to time, different myths were called into question: the cosmic egg, the Brahmanda of the Sanskrit scriptures, the Hiranyagarbha of the Vedic ones, or the primeval sea of Sumerian mythology, all the way through to the more or less anthropomorphic divinities, like those of the Greeks and Romans.

It is interesting to note that one of the standard cosmogonic elements of the pre-Socratic philosophers, such as Hesiod, is Chaos. In Hesiod's *Theogony*, Chaos began to exist before anything else. For Heraclitus, primeval Chaos, located below the Earth but above the infernal abyss of Tartarus, was the true foundation of reality. Thinking about the importance of quantum fluctuations in the vacuum (which we will talk about later) to the birth of the universe, we cannot fail to be impressed by the similarity of these concepts, expressed thousands of years later and in completely different cultural contexts.

With the advent of the scientific revolution, things have changed substantially. In the last century in particular, a number of theoretical and experimental observations and discoveries—in the direction of both the infinitely small and the infinitely large—have given rise to complex scientific reasoning about the birth of the universe. As we will discuss in the following chapters, today we are talking about how a universe originated out of "nothing," characterized by an incredibly violent initial expansion, the Big Bang, which in turn generated the elements, particles, and force fields and out of which the current universe evolved.

As we will discuss in some detail, we have arrived at this description of the first moments of the universe by bringing all of the available data and tools of scientific analysis to bear, information obtained thanks to a level of technological development that has allowed us to expand our observation capacities enormously.

Surprisingly, it was exactly this recent rapid progress that made us understand that the light under which we are searching for the keys of knowledge covers only a tiny part of the universe. That is to say, only the visible matter and forms of energy, a measly 5 percent if we're being generous. As previously mentioned, the remaining 95 percent is pitch dark, formed of matter and energy that, for good reason, we call "dark." Together with Socrates we can say that we know that we don't know. Part of the fascination of contemporary scientific adventure lies in this explicit awareness of that limit.

But let's not anticipate the conclusions: for the moment we would like only to get the big picture, so let's go step by step and go back to the "beginning." One would like to say that the Big Bang was, by definition, the dawn of the cosmos. But it's not that simple.

To begin with, a sunrise follows a sunset, which in turn can occur only if there was a sunrise before it. But what do we do if there was no "before"? What if time emerged precisely from the presence of matter and interacting forces? And what if, due precisely to the lack of time "before" the beginning, the concept of the universe as the effect of something antecedent makes less sense? These are basic questions that contemporary science is asking itself in order to understand what is time in the extreme conditions of the Big Bang.

Second, if by the first dawn we mean the light connected with the Big Bang, we are clearly thinking of watching the evolution of the fireball from afar. And what if there is no "from afar"? What if it were not possible to imagine being "outside" of what the first event was? Perhaps we can imagine ourselves as simply being "inside" a uniformly luminous universe—a very different perspective from a dawn.

Finally: How do we explain the energy needed to create this great expansion? Where does it come from and how is it distributed in the parts of our current universe? We will see that this issue is also holding a radical surprise for us: the possibility that the total energy of the universe could be equal to exactly zero. Beyond the idea of the impossibility of something preceding the birth of the universe, discussed above, there is also the idea that it is not all that necessary to introduce from nothing a quantity of energy that was not there before, an event incompatible, after all, with the known laws of physics. The modern vision of the universe, its origin and its evolution, tends to clearly separate physical reality from reflections on the dimension of transcendence, rescinding the traditional point of contact the origin represented.

All these radical ideas and concepts connected with the description of the initial moments of the universe should suggest to us that, when we reflect on the origins, we must be careful not to get carried away by our vivid but fallacious intuition. Better that we follow a less anthropomorphic and more rigorous line of reasoning. Also because in delving into this discussion, and the uncertainties that accompany it, we will soon find that we are in good company. The issues we are about to address have been at the center of scientific discussion for more than a century, and are still largely the subject of speculation.

A major example are the ideas about the origin of the universe shared by the very same scientists who contributed to the development of the Big Bang model. For instance, in 1931 Einstein was still fascinated by the idea of a stationary universe, which never began and would never end. This was in spite of the fact that his general theory of relativity, developed around 1915, favored the concept of an evolving universe. Einstein then quickly abandoned the stationary universe model—which Hermann Bondi, Thomas Gold, and Fred Hoyle still strongly supported in 1948—but remained dissatisfied with the idea of a universe that changed over time. Einstein's fate was a curious one; more than once he tried to fight the consequences of some of his most profound ideas. He would be the most authoritative critic of quantum theory, despite the fact that he himself had introduced the photon, the elementary quantum of energy. His observations were never trivial; it took half a century to dismantle his critiques of the nonlocality of quantum mechanics. Einstein himself introduced the cosmological constant to counterbalance the effect of gravity in his equations, a concept he later discarded, as we have seen, calling it his "biggest blunder." Half a century later, the discovery of dark energy—the existence of which, as we shall see, results from the observation that the universe's rate of expansion accelerates over large distances—suggests that the cosmological constant plays a fundamental role in the evolution of the universe and that, based on experimental observations acquired in 1998, is greater than zero. A century-old discussion, involving aesthetic intuition, profound mathematical theories, and refined experimental observations: there is little doubt that this debate will further develop in the future.

7

SPACE

I have nothing against Descartes, but I am convinced that we have a score to settle with his famous coordinates. It would seem to be just a simple system for ordering reality; what harm could it do to attach a little mental label, a set of numerical coordinates, to every point in space and, perhaps, even time? Reality is geometric; space coincides with extended matter which can be dominated by mathematics. The system was successful and became well established. In particular, it evolved toward the conception of Newton, who used it to develop his famous dynamics and the description of universal gravitation, based on an absolute space and time, existing independently from matter, not alterable by measurement and observation. Even today, in harmony with Newton, most of us think of a space and time described by Cartesian axes that run without stopping from zero to infinity. Once we've downloaded the GPS app for spacetime, we can navigate toward the Big Bang as if we were taking a trip to Rimini to soak up the Mediterranean sun.

But space contains much more than Newton thought. At the beginning of the twentieth century Einstein arrived, and the questions became more complex. With special relativity, we came to understand that, in the physical world, both space and time can only be defined, through rigorous procedures, as the differences between two physical measurements, which involve clocks, beams of light, and rulers—much more than just mental labels. Nothing is absolute in these quantities; it all depends on the relative speed between two observers. Thanks to the work of Dutch physicist Hendrik Lorentz, the Lithuanian-German mathematician Hermann Minkowski, and the German mathematician and physicist Georg Friedrich Riemann, the effects predicted by Einstein's special relativity were formalized. When comparing the measurements made in reference systems moving at a constant speed, the scale of time expands and that of space contracts. But this is still nothing in comparison to what emerged with general relativity, a theory where gravity becomes the *deus ex machina* that modifies the metric of the coordinates in every point of space and time (we already discussed the metric and its extraordinary properties in chapter 3). Because of the force of gravity caused by the presence of mass, or its equivalent, energy, space, and time are linked by an intimate relationship and become curved, like the wind blowing through the leaves makes them rustle. Curved, you might say, relative to what? Relative to an abstract Cartesian system in which there is no gravity and in which the metric doesn't change between one point and another. Finally, we shouldn't forget quantum mechanics. The vacuum hosts fluctuations of spatial and temporal metrics that characterize the curvature of space-time, due to Heisenberg's uncertainty principle.

At this point the question becomes: If Descartes had been born and lived on a trampoline in constant motion, like the space-time described by Einstein's relativity and quantum mechanics, would he still have described the world with his famous coordinate system? It's not an idle question. Our intuition is formed on the basis of the tools and concepts available to us.

The surface of a sphere is the perfect example of a two-dimensional curved space, which is therefore non-Euclidean. There is nothing straight on its surface and the sum of the internal angles of any triangle drawn on its surface is greater than 180°. Our intuition does not allow us to describe the sphere without immersing ourselves in three-dimensional Cartesian space where the axes are straight lines. Let's put ourselves in the shoes of a two-dimensional being forced to move on the surface of the sphere and not capable of intuiting the existence of a third dimension. It would be hard for this being to realize that the shortest path between two points is an arc of a circle and not a straight line. In fact, the curvature develops in a direction outside the two dimensions in which it lives. For this being, the three-dimensional arc of a circle *is* the shortest path between two points on the sphere. However, if this two-dimensional being were ingenious enough, it could understand whether it was living on a flat or curved surface, for example, by precisely measuring the sum of the angles of the triangles drawn on the sphere. Or, if it were patient, it could start covering the surface on which it lives by painting concentric circular stripes with a brush. If the surface is curved and closes in on itself, initially these stripes would increase in length as each one was added, but, after a while, having reached a maximum length, they would begin to decrease. In the end the being would find itself surrounded by the painted surface; just like a cartoon character who finds it has painted itself into a corner.

When we're dealing with three dimensions, something similar happens. With a little ingenuity, we can understand if this space is curved or not. Borrowing an idea from Stephen Hawking, we can proceed in a way similar to what was just described. To find out whether the volume of a three-dimensional sphere is or is not the surface of a four-dimensional hypersphere, we could begin by spray-painting space, starting from a point and painting ever larger spherical shells. If we are living in curved space, at some point, to our great surprise, we would find that the spherical shells would begin to get smaller and smaller and that eventually we would find ourselves inside an as-yet-unpainted sphere that would surround us on all sides!

This series of considerations helps us understand how deceptive it is for us to think of observing the universe from "outside." When thinking about this, it's much better to place yourself within a volume of spacetime that closes in on itself and which contains all events, including the expansion of the metric.

As American physicist John Archibald Wheeler said, "space-time tells matter how to move; matter tells space-time how to curve." If physical reality is that of curved space, why do we have to continue thinking in terms of straight coordinate axes? Wouldn't the coordinates determined by gravitational interactions, something that is done in some physics theories, be much better adapted to the task? At the moment we find ourselves in an intermediate situation, one in which we use the formalism of curved space-time to describe gravitational dynamics, while for the other fundamental forces we use flat space-time, only deformed by the simple prescriptions of special relativity, and not by the presence of gravity. The unification of the fundamental forces also involves the ability to use the same description of space-time for all physical phenomena.

Current research is moving in this direction. Theoretical physics has introduced powerful topological analysis tools to unify general relativity and quantum physics. It's not a simple thing, because it goes all the way down to the foundations of physics' reasoning. In short, it has opened up a fascinating area of research in which, for now, there are more problems than solutions. The hope is that the theory of quantum gravity will allow us to understand what happens at extreme energies and in the smallest dimensions, typical of the structure of black holes and the origin of the universe. But we're talking about hope, also because the energy scale is so high relative to that which can be explored with accelerators or with observation of the cosmos that we find ourselves practically without indications of an experimental nature.

The great Descartes will forgive me for having taken him to task, but I wrote these lines because I want to draw attention to an aspect that the reader risks considering secondary but which, in truth, is decisive: space, time, and the physical vacuum are anything but trivial. For example, in talking about what happened near the Big Bang, namely, the initial singularity, we become aware that the properties of the vacuum play a decisive role. It's like saying that, in a football match, the properties of the field are fundamental. In fact, they define the game strategies, and the quality of the grass surface has an influence on how the game plays out. The study of the properties of the physical vacuum, a concept closely linked to the very essence of space-time, is an area where theoretical research is continually developing. That it is a difficult and arduous task to explore this terrain is demonstrated by the fact that, for one, we are not even sure that space-time is curved, as Einstein thought, or discrete (which is to say composed of isolated elements), as some contemporary physicists claim—I am thinking specifically of Carlo Rovelli. In fact, we still don't know how our universe came into being.

In any case, we can acknowledge some reasoning based on the known laws of physics and push ourselves as far as the validity of the hypotheses on which they are based will take us. In the first chapters of this book we introduced the necessary ingredients for the discussion and highlighted some of the conceptual traps we risk falling into. Now we are ready to start on our journey toward the infinitely large and the infinitely small, the enormously complex and the absolutely elementary, a journey punctuated with dizzying visions and unexpected perspectives. To guide us we will have the growing conviction that, despite having made great strides, much remains to be discovered. We are only at the beginning, the dawn, of our understanding of the cosmos.

8

TEN INFERNAL MINUTES

In addition to the types of fundamental forces and the mass of elementary particles, we have seen that what we need to focus on to address the discussion of the very first moments of the Big Bang are space, time, and energy.

As mentioned in the previous chapter, we must resist the temptation to think of the moment the universe came into being as a very intense explosion of fireworks, which developed in a very distant and unspecified point in space, that we observe in the same way we would watch a movie. It doesn't work like this. We must get it through our heads that it is misleading to think that we can observe the Big Bang from the "outside."

However, it is important to note that within this volume that closes in on itself, all of the physical quantities are substantially uniform and will remain uniform in all of the universe's evolutionary phases, up to the present day. In the beginning this homogeneity had to do with the energy density and the temperature. Then, as the universe slowly expands, it is the distribution and the average temperature of matter that are the same everywhere. As we come to the present day, after an evolution lasting billions of years, the uniformity has to do with the number of galaxies per unit of volume, but the principle is the same. No place in the universe is special, or privileged; every point within this volume offers the same panorama. There are fluctuations, as we shall see, derived from quantum phenomena, but even those are distributed democratically, with no place or direction preferred over any other.

We have said that for space it doesn't make much sense to think of an "outside"; for time it doesn't make much sense to think of "before," considering that, before the first instant, there was absolutely nothing. And where there is nothing, there are no variations and time simply does not exist. It's like looking at a photograph. Time emerges simultaneously with the Big Bang; after the clapperboard clacks, the most exciting action film of all time begins.

We are now left with two other available ingredients for describing the birth of the universe: energy and space.

To get to the first instant, we can try to project the film of the universe's evolution backward. From Hubble onward, that is to say since the 1930s, we have surprising experimental evidence that the entire universe is expanding in all directions. Let's think about that for a minute; we're talking about something truly extraordinary. It is not the effect of an explosion. In an explosion, everything expands relative to a central point and the density of exploded matter is by no means uniform as the radius increases. Furthermore, the matter tends to be concentrated on a spherical impact front where most of the fragments are found. However, in the case of the universe, what happens is more similar to a cake rising. Each point moves away from the adjacent one, due to a chemical reaction produced by the leavening agent. The result of this elementary phenomenon, which involves only adjacent points, is that the farther away two points of the cake are, the faster they move away from each other.

Perhaps Hubble's observation that the universe expands in every direction should have been described as the Big Cake rather than the Big Bang, but certainly it wouldn't have had the same dramatic effect. Would you rather imagine the universe as a primordial sponge cake or a primeval super explosion? Having clarified this fact, projecting the film of the universe backward, we see that all of the physical quantities involved—density, energy, temperature—become increasingly larger until they tend toward infinity and no longer make physical sense. It is worth noting that as we rewind, the condensation and concentration of the universe occurs at all points simultaneously. Every point is equal to every other one. As the frames roll backward, the energy of the universe is concentrated in an increasingly smaller volume. The initial singularity of the Big Bang corresponds to a "state" of extremely high density and temperature, not related to a precise spatial "position," "extension," or "shape." We move toward a condition in which the known physical laws lose their meaning, as the presuppositions on which they are based are no longer valid. In the vicinity of the singularity we start groping in the dark, not even able to rely on our intuition, which can actually be counterproductive. Therefore, it is essential to reiterate the limits of this discussion. This is, in fact, the only way we can avoid suggesting a mindset that might lead the reader to think that we have understood something we have not understood.

Let us ask the following question in particular: In projecting this film backward, what is the first frame, the smallest moment of which we are able to say something about the early universe? Quantum mechanics gives a clear indication of the moment in which we must stop extrapolating the laws of classical physics, including general relativity, and when we must instead modify them to take into account Heisenberg's uncertainty principle, which we talked about in chapter 4. Thinking about the origins of the universe, we can call upon this principle in two of its formulations. The first is that which states that the product between the uncertainty of the measurement of the spatial position and the momentum must inevitably be greater than a given value tied to a constant named after the German physicist Max Planck. This is not to say that the momentum, a fundamental variable in dynamics, is not conserved; it simply means that the metric of space and the momentum associated with the system under observation start to fluctuate randomly as we approach the first instant. The uncertainty principle predicts that when the characteristic dimensions of the universe, in the specific case those represented by the curvature radius of space-time, become very small, we enter into a regime of quantum fluctuations. Another pair of variables that obey the uncertainty principle are the uncertainty in the value of energy and of the corresponding time interval in which the measurement of that energy takes place. As we approach very small intervals of time correspondingly large energy fluctuations arise, and vice versa. This is not to say that energy is not conserved in these conditions, but simply that the metric of time and energy associated with the system under observation starts to fluctuate.

The phase dominated by quantum fluctuations is called the Planck epoch, and is equal to times less than 10^{-43} seconds. We are not able to analyze what happens for times less than this, as we do not currently have access to theories that encompass gravity, relativity, and quantum mechanics in a coherent way.

However, energy fluctuations have the characteristic of being able to vary for positive and negative values, and this offers us an interesting opportunity. On the one hand, we have the positive energy corresponding to mass and to kinetic energy, linked to the speed of the particles with mass and photons. On the other, we have the negative energy linked to the value of gravitational potential and the curvature of space-time. This paves the way for an incredible opportunity. It is possible that the energy fluctuations can be very large but of opposite values, adding up to zero so that they conserve the available energy at any given moment; consequently, due to the uncertainty principle, the duration of these fluctuations can become as long as desired. In other words, the uncertainty principle, associated with the conservation of energy, can be used to create a universe from a zero-energy state.

It is therefore possible that the universe arose out of these very quantum fluctuations. Analogous to Planck time, we can define a length equal to Planck time multiplied by the speed of light. If we calculate the energy fluctuations within a Planck volume (a Plank length cubed), and transform them into mass, using the mass-energy equivalence of special relativity, we obtain an exceptionally high density, equal to 10^{97} kg/m^3. We're talking about a mass equivalent to twenty human cells, but compressed into a microscopic volume: a very small mass/energy for a universe, but incredibly dense, enclosed in a microscopic volume in which space-time is strongly curved by the negative fluctuations of gravitational potential. Just for comparison, the density in the atomic nucleus of a neutron star is *only* 10^{18} kg/m^3.

Where does the energy needed to create this hyper-dense state come from? We hinted at it just above: physics holds a big surprise for us, first suggested by Edward Tryon in 1973.

Thanks to the fact that the sum of the positive and negative energy of the quantum fluctuations is always conserved, the total energy of this microscopic universe seething with quantum fluctuations is always zero.

In this way the universe would emerge from a zero-sum quantum fluctuation, the effects of which continue to this day. A truly bold idea, and yet perfectly in accordance with the principles of quantum mechanics. Alan Guth, one of the fathers of contemporary cosmology, defined it very effectively as the "ultimate free lunch."

Let's pause for a moment to think about what we just discussed. It is as if we had a bank, which did not contain a single dollar. But it has a lot of clients, some depositing money, and others borrowing it. Leaving aside the effect of interest, the sum of money that the bank has always remains equal to zero, but in the meantime the economy works. There is no reason why the sum of the money owned by the bank has to have positive value for the mechanism to produce its effects. Another example that might be useful would be that of transforming a flat area into a series of hills and valleys. Overall, the potential energy hasn't changed, but the landscape is completely different. Starting from the quantum fluctuations of the vacuum, the laws of physics allow the activation of a similar mechanism, starting from a state of energy equal to zero.

Let's leave Planck time and the complexity of quantum fluctuations in space-time behind. Instead, let's take a look at what happens in the following stages. Between 10^{-43} and 10^{-36} seconds, the universe expands and cools, entering the grand unification epoch, marked by the unification of the fundamental forces. In this period, the temperature is still too high to allow either the differentiation of the various types of particles or the independent action of the four fundamental forces that we know. There is only one super force, which acts symmetrically on everything. It's like being in a big crowd, everyone crushed against each other; it is impossible to tell who is heavier and who is lighter: the energy is so high that the mass, different for each particle, cannot yet play a meaningful role. At the end of this epoch, as the universe expands and the temperature drops, the super force gradually breaks down into the single forces that begin to have different intensities and characteristics. The first to break away is gravity, leaving the electronuclear force to determine the dynamics of this period, during which the temperature drops to less than 10^{28} Kelvin. At that point the electroweak and the strong forces separate and we now have three of the four forces that we observe at work even today. Currently, the strength between gravity and electromagnetic interactions varies by forty orders of magnitude. It is no small matter when the forces separate from each other: it changes the characteristics and structure of the physical vacuum. To go back to the example of the football field, it's as if, due to a violent hailstorm, the characteristics of the turf change while the game is in progress: two completely different styles of football are played before and after.

In the case of the electronuclear force transition into electroweak and nuclear or strong forces, things apparently become complicated: the transition doesn't produce a stable vacuum, but rather a metastable state. It's a bit like when water is supercooled without freezing: you get an unstable condition that, as soon as it is disturbed, instantaneously transitions the entire volume of water from liquid to solid, with a violent emission of heat.

Don't play around with the vacuum!

The American Alan Guth and the Russians Andrei Linde and Alexei Starobinsky were the first to realize how, taking the general gravity equations into account, the metastable situation of the physical vacuum produces enormous pressure on the vacuum itself, which overcomes the attractive effect of gravity and causes the metric—the property of space that defines the benchmark for cosmic distances we discussed in the third chapter—to violently expand in all directions, literally exploding. This process of expanding the metric has been called *inflation*: during a microscopic fraction of a second, the metric expands at a speed well beyond the speed of light, causing distances to increase by an enormous factor, approximately 10^{26} times. The metastable vacuum that was created in the initial instants starts to occupy an enormous volume; the energy of the universe is redistributed by being absorbed by the rapidly expanding vacuum; and the curvature of the space-time metric reduces, consequently decreasing the density of negative energy. Energy is conserved at all times, but is redistributed in a universe that becomes vastly larger in an incredibly short time. We have seen that special relativity's maximum velocity limit does not apply to the expansion of the metric.

To understand the significance of the change of the metric, we can go back to the example of the football field. Changing the rules by increasing the metric of the field can be done instantaneously; however, then we find ourselves with players placed far apart from each other and forced to run much larger distances. At the end of the inflationary expansion, the universe was vastly less dense, exceptionally cold, and, above all, much more uniform. Parts of the universe that were close at first have moved apart so quickly that they have gone beyond the mutual horizon guaranteed by the speed of light, thus losing any causal relationship. The violent quantum fluctuations that characterized the previous phases of the universe are now just a memory, fading without completely disappearing. The energy accumulated in the vacuum during the inflationary process is finally released when the metastable vacuum decays into a stable state. In this next phase of the universe, now much larger and much colder, it rapidly heats up again, creating particles with an exceedingly high energies and temperatures. In a sense, the mechanism of the Big Bang, in terms of a great primeval explosion, begins with the end of the inflationary process, even if we have seen that the expansion we are talking about is uniform and does not correspond to an explosion. The energy passes from the vacuum to particles with mass and kinetic energy. The sum of the energy—both the positive linked to the particles and the cosmological constant, and the negative linked to gravitation and accumulated by the curvature of space-time—is still zero, then as now. The inflationary evolution of the vacuum has therefore allowed the transition between the initial microscopic universe and the much larger one in which there are particles and force fields. However, remember that in all phases of the universe, even during and after inflation, the fundamental quantities of the initial state that corresponded to the physical vacuum are conserved: energy, momentum, angular momentum, and electric charge are equal to zero. Not only that, but up to a certain time in the early evolution, symmetry between matter and antimatter should also have been respected (we'll talk about that in chapter 27), ensuring that particles and antiparticles are produced in equal number, guaranteeing that the total sum is zero.

The post-inflationary expansion then continues for about 9 billion years, at a much slower and more regular rate. The speed at which any two points in the universe move away from each other is fixed by a quantity called the Hubble constant, which corresponds to a recession speed of around 70 kilometers per second for every 3.26 million light-years of distance. Based on what has been established in the preceding paragraphs, the further the distance between two points, the greater the relative speed of recession. This expansion rate started to speed up again about 5 billion years ago, possibly due to the activation of an additional supply of repulsive energy, known as dark energy, present in the vacuum. It is such a weak supply that it becomes observable only when the distances between objects are cosmological, that is to say, comparable to the current size of the universe. Although almost 14 billion years have passed since the first instant, the properties of the vacuum continue to influence the evolution of the universe!

Continuing with our story, between 10^{-36} and 10^{-32} seconds is the electroweak epoch, dominated by a force that combines electromagnetism and weak interactions, mediated by the massless photon and the heavy intermediate Z and W particles, respectively. Here we begin to approach the energy scales that can be studied in the collisions at the CERN accelerators, the modern microscopes for the study of the microcosm, which we will discuss later. In fact, the neutral Z^0 boson (a particle about ninety times more massive than a proton, which allows interactions between neutrinos and other neutral particles) as well as those charged $W^{\pm}$ (particles about eighty times more massive than a proton, which allow interaction between neutrinos and other charged particles) were discovered at the CERN accelerators in 1983. Together with the photon, intermediate electroweak bosons carry electroweak force between the various interacting particles.

Much later, considering the timescale used so far, we enter the quark epoch. Quarks are the elementary constituents that form hadrons, namely, particles such as protons and neutrons. Around 10^{-12} seconds, the weak force separates from the electromagnetic force and the elementary particles acquire a well-defined mass, interacting with the Higgs field, another fundamental particle discovered at CERN in 2012. Now the universe has become a piping hot soup of elementary particles and mediators of the fundamental forces, which are all acting independently. These conditions can now be reproduced in high-energy particle accelerators such as the LHC at CERN. The temperature is still too high (around 10^{15} Kelvin) to form hadrons, the bound states of quarks. At this stage the universe is, likely, still symmetrical between matter and antimatter, and also around 10^{-12} seconds should mark the beginning of baryogenesis, the gigantic annihilation between particles and antiparticles at the end of which only a fraction of one part per billion of the initial matter manages to survive, thus gaining the upper hand. Finally, at around 10^{-5} seconds the hadron epoch begins. Quarks bind together to form protons, neutrons, pions, and other particles. From that moment on, the strong force remains confined within these nuclear particles, the building blocks of the atomic nucleus.

About a second after the Big Bang, the lepton epoch begins. Leptons are elementary particles like the electron and the neutrino. At a temperature of around 1 billion Kelvin, the neutrinos separate from the rest of matter, now lacking the density to interact with them, going on to form the cosmic neutrino background. In the next ten seconds the photon-dominated epoch begins. Electrons and positrons annihilate each other and only a small fraction of electrons remains. Together with the protons they form a hot plasma, opaque to radiation, a sort of gigantic, expanding star that fills the entire universe. In the three minutes that follow, primordial nucleosynthesis takes place, creating nuclei as well as the lightest isotopes: deuterium, helium-3 and helium-4, along with traces of lithium. Around the tenth minute, primordial nucleosynthesis ceases as the density and temperature of the plasma is no longer sufficient to sustain the nuclear fusion reactions.

We have seen how the first ten minutes after the Big Bang were marked by a frenetic succession of conditions of matter so extreme that only a small part of them are reproducible in the largest modern particle accelerators. From this moment on, the rhythm changes drastically and everything starts to develop more slowly.

9

AND THERE WAS NOT LIGHT

For the 379,000 years that followed, the universe continued to expand regularly: filled with a high-temperature plasma formed by electrons, protons, helium, traces of lightweight nuclei, and photons. At this point, it is also saturated with neutrinos, but these hardly interact at all with the rest of the matter. The metric continues to expand at a constant rate, moving each point away from the others, reducing the density of the plasma and neutrinos, and consequently their temperatures. It is this continuous, unstoppable expansion that causes the dimensions of the universe, at every moment of its evolution, to be much larger than its age multiplied by the speed of light.

To understand this effect even better, let's think about a runner competing in a 100-meter race: at the halfway point, the track extends and becomes twice as long. The runner keeps going but must cover another 100 meters to get to the finish line. In total, the athlete has run for 150 meters on a track that is now 200 meters long. This is why, at this point in its existence, the universe accessible to us is larger than the speed of light multiplied by its age, which makes it about 46 billion light-years. It is important for us to remember that there is no "outside." All of the matter and energy that form the universe at every instant are distributed throughout the entirety of its volume, which is expanding at a speed dependent on the distance between the various points. If we think about it in two dimensions, it's a little bit like we are inflating a balloon and observing its surface. Under these conditions we should not be surprised that, between two sufficiently distant points, the reciprocal speed of separation can exceed that of light, making these two points causally disconnected, starting from a certain instant in the life of the universe.

What's so special about the 379,000-year mark? Before this time, the temperature and therefore the energy of the electrons and protons was too great and that meant these oppositely charged particles could not yet bind to form hydrogen atoms, just as water vapor, under normal conditions, cannot condense above 100°C. Even if at this stage the temperature of the universe is around 4,000 Kelvin, only a little less than the surface temperature of a star like our Sun, we must not confuse this state of the universe with that of a giant star. In this young universe, plasma density is very low—about 330 million electrons per cubic meter. We're talking about a density that is over a billion times higher than that of our current universe, a quarter of an electron per cubic meter, but still 1,000 billion less than that of the Sun. In such low-density plasma a photon bumps into an electron after having traveled an average of 5,000 light-years. This dimension may seem vast but I assure you that, on a cosmological scale, it isn't.

This is why it is reasonable to imagine that, before recombination, the universe was opaque or, better, filled with a luminescent mist. Once we drop below the recombination temperature, and plasma is replaced by neutral atoms, the length that photons could travel freely vastly increases and the universe becomes transparent to light.

Therefore, at some point, there was *not* light.

Like a breeze blowing fog out of a valley, the plasma, luminescent from the constant rate of collisions, starts to form a gas of neutral atoms— mostly hydrogen, but also helium and traces of other lightweight elements. The photons start to travel freely, and after a while, the universe falls into a state of pitch blackness, in the sense that photons and neutral matter stop interacting. This period lasts for a long time, several hundred million years, during which gravity, slowly but inexorably, exerts its attractive force. We talked about the violent initial fluctuations and their enormous dilution during inflation. Diluted yes, but not gone. These fluctuations correspond to more or less dense areas of matter, at higher or lower temperatures. The difference in temperature between one area and another is minimal; we are talking about parts per million, yet enough to break the symmetry locally. The colder, denser areas start to attract atoms from adjacent areas. As they collide, excess energy is dissipated then removed via photon emission, and increasingly larger aggregates of condensed matter form. At a certain point, after hundreds of millions of years, so much mass has collected at any one point in the universe that an extraordinary phenomenon, the *ignition* of the first star, takes place. The nuclear fusion processes taking place inside its core, compressed by the weight of the surrounding mass, release so much energy that the star reaches a temperature of a few thousand Kelvin and begins to emit light. Like a match in the dark, the first star illuminates a universe that became dark hundreds of millions of years earlier: now the universe has grown a great deal and begun to light up.

10

THE FIRST DAWN OF THE COSMOS

Perhaps this is really the first dawn of the cosmos, the ignition of the first star in the universe. An extraordinary phenomenon, seen by no one, but certainly observable, unlike the initial expansion that gave rise to everything but where the separation of light and dark was undefinable. After the first star, the second ignited, and so on. Like at a rock concert, where people light their lighters or cellphones, one after the other, until there are thousands—in the case of the universe millions, billions.

But how do stars work? It is a very delicate and astonishing balancing act between gravity's attraction and the rhythm of nuclear fusion reactions, which produce energy flows that offset its pressure. Let's consider the rather unique nuclear properties of elements that exist in nature: looking at increasing mass values, starting with hydrogen and ending with iron, we can see that there is an energy advantage in forming heavier atomic nuclei by combining lighter nuclei. As soon as conditions allow, as we see with stellar cores, the nuclear *fusion* reactions among lighter nuclei tend to produce heavier nuclei, releasing energy. This process is very similar to exothermic chemical reactions. The water molecule is more stable than oxygen and hydrogen separately. In the right conditions, for example, in the presence of a spark, water is formed, releasing energy, often with an explosion. On the other hand, in the case of nuclei heavier than iron it is easier to divide an atomic nucleus than to add a piece to make it bigger; this is the physical basis of nuclear power plants or atomic bombs, which generate energy through the *fission* (splitting) of uranium and plutonium.

But what are the right conditions for fusion? The fact is that, in order to fuse two light atomic nuclei, it is necessary to overcome what is known as the Coulomb force, that is, the very strong electric repulsion between positive charges close together. If this force is overcome, even for an instant, the nuclei are attracted by the nuclear force, which, at distances equal to the diameter of the atomic nucleus, is much stronger than the electrostatic force.

The first stars are made up of the lightest elements: hydrogen and helium. These quickly become very large, given the abundance of available atoms in the immense initial nebula. As the mass of the future star grows, the gravitational force increases and compresses the stellar core, making it dense to the point of risking gravitational collapse, the condition in which atomic matter becomes a clump of neutrons or even a black hole, which we will discuss later. But it is this same gravitational pressure that, in the center of the star, makes nuclear fusion reactions possible, producing heavier elements and, at the same time, releasing an immense amount of energy.

This is the energy that, in the form of pressure from the flow of photons that move toward the surface to be radiated, maintains the star's equilibrium. Inside the stellar core, there is the ongoing nucleosynthesis of increasingly heavier elements—a process that creates new atomic nuclei from already existing protons, neutrons, and other nuclei through a complex series of nuclear reactions, some faster and some slower. At some point in its life cycle, the star runs out of the light elements that have been its nuclear fuel, and the structure loses its equilibrium. The core then implodes under the pressure of the force of gravity in ways that are determined by the size of the original star. Smaller stars have a less violent evolution: first, their volume increases and they become red giants, then they collapse into white dwarfs surrounded by protoplanetary matter. Is this important? Yes, because this is what is destined to happen to the Sun in about 5–7 billion years. For larger stars, the explosion is very violent. The core collapses into a neutron star or a black hole, while the rest of the star explodes, creating a supernova, distributing most of its stellar matter, now enriched with heavier nuclei, into the surrounding space.

There is one more thing to say about smaller stars: they live much longer than larger ones. They are perfect for forming planetary systems and feeding them with a fixed amount of energy over longer periods of time, ensuring the necessary conditions for the establishment of planetary climates and, possibly, for the development of life. On the contrary, the largest stars are shorter lived: a star that is twenty times heavier than the Sun will have a lifetime a thousand times shorter, about 10 million years. The role of large stars is that of nuclear forges for the formation of atomic nuclei up to iron, which would otherwise not be present in the cosmos. In addition to this, they are necessary for the creation of black holes of various sizes, indispensable in the creation of galaxies, which are almost always organized around a large, central black hole. In short, every type of star has a role in the development of the universe, or, if we prefer, the universe has become what it is precisely because of the properties of the stars of which it is composed, fundamental elements in the transformation of the neutral, inert gases with which it was filled after the recombination into the complexity of the current universe.

The hundreds and millions of years that followed the recombination phase witnessed the formation of countless stars, followed by the explosion of larger ones. This sequence has been repeated many times, each time with new stars characterized by heavier elements in their composition; we therefore speak of stars that are more or less *metallic*. Attracted by enormous black holes, these stars led to the formation of large structures that then evolved into the galaxies that still fill the observable universe.

11

A SPECIAL STAR

There is a star about which we care a great deal, the Sun. We have seen how stars are formed inside molecular clouds. In all likelihood, the nebula that produced the Sun also produced thousands of other stars, a true stellar nursery, like the Orion nebula, which is about 1,200 light-years from Earth and is even visible to the naked eye in the Milky Way, illuminated by the stars that have formed inside it. The history of the Sun and the solar system is one that began about 4.5 billion years ago, when this area of the cosmos was part of a huge molecular cloud that extended across at least 65 light-years.

As discussed in the previous chapter, this cloud was the product of generations of stars that formed and exploded in this region of the galaxy. Ninety-eight percent of it was composed of molecular hydrogen, helium, and traces of lithium, resulting from the phases following the Big Bang. The remaining 2 percent was made up of heavier elements, produced in the nucleosynthesis of the previously existing stars. It is precisely the heavier elements that help us to understand the sequence of events: the unstable isotopes of the heavy nuclei produced inside the stars—such as iron-60 (^{60}Fe), which has a half-life of 1.5 million years—are an authentic cosmic clock. For example, the analysis of the most ancient meteorites allows us to date, with some precision, when the last stellar explosion occurred—the one that probably generated the shock wave that started the formation of the Sun and the coalescence of this part of the nebula. Then, the events that led to the formation of the solar system followed one after the other in rapid succession, that is to say, "rapid" on a cosmic scale: over the course of 100,000 years, a rotating, protoplanetary disk formed, which then separated from the rest of the nebula. It is initially about 200 astronomical units (AU) across, which is about twice the size of the current solar system. Within it, the dust particles collided with increasing frequency, gradually forming larger granules, called planetesimals. At the center of the rotating disk the growth rate is faster and the temperature is higher. It is here that the Sun formed over the course of about 50 million years. In its first phase, we are talking about a protostar that grows by attracting material from the nebula without yet triggering the nuclear fusion process. The increasing force of gravity then leads to a collapse that, over the course of 500,000 years, causes it to transition from a protostar to a G-type main sequence star (stars are classified into nine classes according to their characteristics, which corresponds to a specific type of evolution), which derives its energy primarily from a sequence of reactions that lead to the fusion of hydrogen nuclei that form a helium core in addition to releasing a certain amount of energy.

It is worth analyzing this sequence carefully. The first step is the creation of deuterium, ^{2}H, a stable nucleus formed by one proton and one neutron. The number in the exponent to the left of the element symbol describes its atomic weight: ^{2}H stands for a hydrogen nucleus with two nucleons (i.e., a proton or a neutron), not to be confused with H_2, which as we know indicates a molecule formed by two hydrogen atoms, each with a nucleus formed by only one proton.

The production of deuterium occurs through the fusion of a pair of protons and the simultaneous transformation of one of the two protons into a neutron, through the emission of an electron neutrino and a positron (the antiparticle of the electron), which quickly annihilates itself with an electron, producing energy. The deuterium nucleus subsequently fuses with one of the many remaining free protons, forming ^{3}He, a nucleus of the helium isotope made up of two protons and one neutron, in addition to the emission of a gamma ray, a quantum of highly energetic light. Two ^{3}He can then fuse, creating a nucleus of ^{6}Be, a beryllium isotope; after just 5×10^{-21} seconds, this unstable isotope decays into two protons and one helium nucleus, ^{4}He. At the end of this series of reactions we find ourselves with four stable particles, of which one, helium, was not present at the beginning, in addition to a certain amount of energy that heats the solar plasma. Of course, in the furnace of the Sun's core, these reactions occur randomly, along with many others. But this is the sequence that provides much of the energy emitted by our star. The ^{4}He nucleus has a total mass smaller by about 0.7 percent of the mass of the four protons that participated in its formation. It is the energy corresponding to this mass that has been transformed into heat, simultaneously creating a more stable, nuclear bonded state. Although not a particularly large star, the Sun's numbers are still impressive: every second, 600 million tons of hydrogen are fused into ^{4}He. In terms of relativistic transformation of mass into energy, it means that 4 tons of mass are consumed every second. For a more comprehensible comparison, we can say that, so far, the Sun has converted ten times the Earth's mass into energy, which is equivalent to—another significant fact—about 0.03 percent of its mass.

The heat generated in the Sun's core takes up to about 170,000 years to come to the surface and be radiated into the surrounding space. This is a very long time, over the course of which, through an infinity of collisions with plasma, the photons work their way toward the exterior. The neutrinos are much faster; as they interact much less with the plasma, they work their way through in seconds. The conditions of equilibrium between gravity and the flow of heat generated by nuclear fusions correspond to a surface temperature of about 5,770 Kelvin and an internal temperature that, in the center, reaches 15 million Kelvin.

We know how old the Sun is, but do we also know how long it will keep going? At this moment, taking into consideration the class to which it belongs, we can say that it's a middle-aged star. Its brightness is fairly stable; over the last 3 billion years it has increased by 20 percent and the same will happen over the next 2 billion years. Considering that these variations have occurred over such a long period of time, it represents a variation of less than 1 percent over the last 100 million years, fully compatible with human survival. We also have to take into account that, with an eleven-year cycle, the brightness of the Sun periodically changes by about one part per thousand, in proportion a much greater change than that of average duration, but in any case, always limited. Being a small star, the Sun is particularly adapted to ensuring stable irradiation conditions, both for the delicate phase of planetary formation and for time intervals comparable with the duration of the evolutionary processes of life as we know it on Earth.

What will happen to the Sun? In about 5 billion years, the end of its long, stable phase, it will become a red giant when the protons that feed the core are exhausted and other nuclear reactions become dominant. The equilibrium within the star will be completely altered, and the outer part will swell up and engulf all of the inner planets, up to the Earth. After a complex series of radical transformations, it will become a white dwarf with a mass equal to about half of its current mass, and continue to burn for a few billion years with whatever remains of the planetary system after the red giant phase. This is a common fate for this class of stars, a sequence of events that has happened and will happen billions of times in the universe, but which is of unique importance in this case, because *this* star is the one to which we owe our existence.

12

THE DAWN OF THE SOLAR SYSTEM

While the Sun was forming, a similar process of gas condensation and gravitational coalescence of dust was happening in various parts of the protoplanetary disk. The delicate balance of this process requires very precise thermal conditions. We understand why it is so important that the central star is in a condition of stable radiation: overly high temperatures, induced by excessive variation in solar radiation, would have prevented condensation, frustrating the weak action of gravitational attraction between the small masses that were forming. It is probable that, at a certain point, there were hundreds of protoplanets orbiting around the Sun; as their mass gradually increased, so did the relevance of the gravitational effect while that of solar radiation diminished. Over time, the planetesimals and protoplanets merged with each other, forming ever larger bodies that culminated in the formation of the current planets.

Also due to thermal equilibrium, the composition of the planets is very different, depending on how far they are from the Sun. Once again, solar radiation plays a decisive role: in the early stages, radiation pushes the lighter gases—those containing hydrogen, like water, ammonia, and methane—toward the outside of the protoplanetary disk, increasing the percentage of heavy elements on the inside. These elements condense at higher temperatures: in the form of dust and granules, these nuclei tend to increase in size when they collide with each other. Beyond a certain

distance from the central star, called the *frost line*, the ambient temperature drops below 150 Kelvin, about 130 degrees below zero. In these conditions even lightweight materials condense, forming increasingly large
clumps. The current frost line is found at a distance of about 2.7 AU,
between Mars, the last of the small rocky planets, and Jupiter, the first
of the gas giants. Not surprisingly, that is the same distance at which we
find the main asteroid belt, testimony to a planetary formation process
that has stopped. Most likely this was due to Jupiter. The rapid formation
of the gas giant that, if it had reached a radius of about four times larger,
could also have become a star, disturbed the process, emptying the belt
and attracting most of the asteroids that formed it. Today the asteroid belt
contains about 4 percent of the Moon's mass, less than one-thousandth
of its initial mass. I am particularly fond of the main belt in that one of
the asteroids, 21256-1996CK, to be exact, was recently christened 21256
Robertobattiston: I imagine it like the planet of Saint-Exupéry's *The Little
Prince*, with a splendid view of the solar system.

Explaining the formation of the gaseous planets farther from the Sun
is a more complex task. It is very likely that they formed closer to the Sun
and then migrated toward the periphery due to gravitational interactions
with the gas giants Jupiter and Saturn. As was to be expected, there was
never a first dawn in the solar system, just a long sequence of processes that
led to the formation of the system in which we live. Slow processes for
us, as we know, but relatively rapid if we take into consideration the timescale of the cosmos. The interesting thing to point out is that these processes are still ongoing. In short, the solar system, the Sun, and the Earth
are not stable; they are simply evolving, even if this is taking place over
periods of time immeasurably longer than those with which we mark
human events.

In analyzing the formation of the solar system, for example, it strikes
us that the scales of time and space we are dealing with are vastly different from one another. The initial nebula, that is, the soup of atoms, molecules, and dust with which we started at the beginning of this chapter,
is the result of billions of years of activity in which the nuclear forges
within the cores of relatively short-lived stars transformed the nuclei
formed in the Big Bang—hydrogen, helium, and traces of lithium—into
increasingly heavier nuclei, up to and including metals. From a spatial

perspective, this nebula extended over an area tens of thousands of times larger than the solar system. The explosion of each one of these stars, at the end of its life span, dispersed the material of which it was composed, which then formed other stars, enriched with heavier elements, which in turn exploded and dispersed their content. The Sun is made of this mixture of elements. As already mentioned, it took 50 million years to form, but only half a million years to become a stable star, following the activation of internal nuclear fusion reactions. Four and a half billion years ago, the ignition of our star catalyzed the formation of planets. Over the course of a few million years, radiation allowed the separation of lighter elements from the heavier ones, resulting in the formation of the small rocky planets and the large gas giants. What was not gathered in by the planets or the asteroids was quickly swept away by the radiation, bringing an end to planetary formation. In the period following the Sun's ignition, gravity played a defining role both in quickly forming the existing planets and their satellites and, at least in the case of the main asteroid belt, in preventing their formation.

Let us pause for a moment to admire this extraordinary orchestral symphony created by the chaotic interaction of different elements and forces, taking place in incommensurable periods of time and conditions, which eventually generated the solar system, within which lies the Earth—a planetary system that, to us, appears majestic, orderly, eternal. But we know that it is the temporary result of interactions without intentions, of mechanical laws at work on different scales, with each physical event and each material element oblivious to all the rest, a system that seems ordered to us in that we are observing it at a very particular moment in its evolution. It is as if we are looking at a high school graduation photo, forgetting that the young person in the photo, like all people, was born in blood and will turn to dust. The idea that this is a special moment in the solar system, that it has evolved from chaos into order on purpose and just for us, like all of the rest of the universe, is hard to resist. However, physics does not support us in this illusion. The second principle of thermodynamics tells us that all of the processes in which heat is dissipated—from the clapping of hands to the internal combustion engine—inevitably increase the disorder of the universe. This is why the universe as a whole can only tend toward maximum disorder, which is measured by *entropy*. So, the evolution from the protoplanetary nebula to the current order of our solar system has produced much more disorder than there was at the beginning. This disorder was discharged into the infinity of photons produced during the various stages of its evolution; consequently, the total entropy value has increased. It is something like what happened to the initial energy, which maintained a zero sum but was redistributed between different parts of the universe. In the case of entropy, some parts have achieved greater order at the expense of others, which have become more disordered. However, in this case it was not a zero-sum game. The entropy of the universe increases inexorably over time; in fact, the entropy value is probably the most correct way of defining the age of the universe.

Now, let's get back to our young solar system. For a few hundred million years, violent encounters between asteroids and planets continued, like that titanic one that likely resulted in the creation of the Moon. From a planetary point of view, the structure of the solar system can be considered to have remained "stable" for more than 4 billion years—a very long time. So, imagine how many earthly sunrises and sunsets there may have been of which we do not and will never have a memory.

Asteroid collision is also a phenomenon that we deal with only briefly because it happens on such extended timescales that it is unlikely that there will be a major impact in the short run. These collisions, in fact, did not end with the formation of the planets; they just reduced in intensity and frequency. It is important to understand this phenomenon because it has a bearing on our survival. Naturally, from time to time, one of the millions of asteroids in the main belt is disturbed by some gravitational effect and begins to waver along in a chaotic orbit, even intersecting with those of other planets. There is a risk that, at some point, one of these may collide with Earth. Is it possible to protect ourselves from an asteroid that has decided to impact our planet? Honestly, we're not very prepared for the event, but we're getting the tools together. First, by scanning the sky to identify any new asteroids approaching our planet, determining their chaotic trajectory with the greatest possible accuracy, and evaluating the probability and time frame of a potential impact. For years the sky has been monitored with telescopes designed to gather just this type of information, maximizing observational sensitivity. Once the trajectories of new asteroids moving between one planetary orbit and another are determined, we are faced with the problem of what to do if the assessment indicates a concrete risk of collision. At the moment we have no technologies that can destroy or deflect objects of this nature, even if planetary protection is an active area of space research. Today the epilogue of the film *Armageddon*, in which heroic astronauts manage to break and deflect the monster asteroid that is heading towards the Earth, is still pure science fiction. In any case, we must learn to live with these kinds of hazards coming from the cosmos, which are characterized by low probabilities but which would have incalculable effects. Like gambling, big wins are rare and come unexpectedly; in this case it would be a major loss! Remember what happened to the dinosaurs? Their disappearance, which occurred about 66 million years ago, was most likely due to the catastrophic impact of an asteroid with a diameter of approximately 10–15 kilometers that landed in the Gulf of Mexico. It happened at the end of the Cretaceous period, and caused the fifth mass extinction, the last one due to natural causes before the ongoing one, which is due to human activity. But we'll discuss this in the next chapters.

13

THE EARTH AND ITS CLIMATE

As everyone knows, even those who are not experts in physics, the Sun is the primary source of our planet's energy. We have seen that the Sun's energy flow varies by less than 1 percent every 100 million years. This is no minor detail. Suffice it to say that, compared with the stability of our star, the response of the Earth to solar radiation has varied in a much more significant way. Consequently, the Earth's environmental and climatic conditions have changed substantially several times over the course of the past hundreds of millions of years, mainly due to endogenous mechanisms.

Moreover, the planet's climate is the result of the complex interactions of multiple factors and mechanisms that are often operating on different timescales even if all of them are in some way connected to the balance of energy the planet receives from the Sun, that is, the difference between the energy received and that which is reemitted or reflected. The Sun is a light-emitting source with a surface temperature of 5,770 Kelvin. The spectrum that reaches the Earth peaks in the range of light visible to us (44 percent), but has a significant infrared component (53 percent) and a small ultraviolet component (3 percent), which our senses are not able to perceive.

What happens to this energy once it reaches the Earth? Thirty-five percent of it is reflected back into space, 51 percent is absorbed by the surface—which reemits 17 percent into space and 34 percent into the atmosphere—and 14 percent is absorbed by the atmosphere, which however reemits 48 percent of that back into space, also irradiating the part received from the surface. Any factor that modifies the radiative properties of the atmosphere, or the surface of the Earth, alters the energy balance. Think of the increase in the density of greenhouse gases (CO_2, methane) in the atmosphere influenced, for example, by volcanic activity or changes in the reflectivity of the Earth's surface generated when there is an alteration in the areas covered by ice or snow, water or vegetation, or, finally, the average cloud cover. It should also be mentioned that changes of this kind have always occurred, periodically, over the course of the Earth's existence, and some of those have even had profound consequences for the climate. Other factors that can influence the balance of energy involve variability in the inclination of the Earth's axis, which has a periodicity of about 41,000 years, changes in the magnetic field, and the temporary inversion of the magnetic poles.

Increasingly precise and in-depth paleoclimatology studies allow us to trace the planet's average temperature trend over the last 500 million years. Based on what we have just explained, it should come as no surprise that, over the course of hundreds of millions of years, the Earth's average temperature has fluctuated around the current value by about 20 degrees. This has caused radical variations in the climate, resulted in the extinction of many of the living species present on the planet, and profoundly influenced their geographical distribution.

On the contrary, what is notable about the Pliocene and the Pleistocene epochs is the greater stability of the average temperature, that is to say, over the 6 million years during which the evolution of the human species took place. In this period the average temperature dropped to about 4 degrees lower than it is now and then stabilized at 2–4 degrees lower, with short peaks above. In the Holocene epoch, the last 11,000 years in which all civilizations have developed, the average temperature has been incredibly stable, within plus or minus 1 degree of what it is today. If it is true that in the course of our evolution, humankind has endured changes in the climate due to a variety of geophysical mechanisms, it is equally true that in the last century we have witnessed something never seen before: the industrial revolution and human population growth have begun to leave a clear imprint on the climate. Greenhouse gas emissions, in particular CO_2, have increased by 35 percent, reaching their highest levels in 420,000 years. The most recent and reliable studies show that this growth is the basis of the increase in average temperature, which has gone up by about 0.8 degrees in 140 years. The serious and looming problem is that it is now increasing at a rate of 0.15–0.20 degrees every 10 years. At this rate, it could increase by 2–6 degrees by the end of the twenty-first century. Remember the timescale? In geological terms, two centuries are the blink of an eye. The fact is that such a sudden change has never been observed. It would be the highest temperature reached over the last 10,000 years, even higher than at the end of the Pliocene (3 million years ago), an epoch in which the sea level was, in all probability, 25 meters higher than it is today. Our existence is based on natural phenomena that have endured over billions of years, but also on phenomena that are measured in millennia or even, as is the case with climate change, in centuries. Our ecosystem develops within a cosmic balance that has allowed us to exist only because the timescales of natural changes are slow; the impact of humankind on the climate has accelerated them precipitously.

Our species' existence in this universe is like that of a tightrope walker who has climbed onto the shoulders of another tightrope walker who in turn is seated on a chair, balanced on a wire, overhanging the void. This miracle of balance can work, of course, but only if the acrobats are well trained and move with the utmost attention, fully aware of the effects of their actions. Will we be able to be that good?

14

THE DAWN OF LIFE

Let us now leave the inanimate universe behind for a few chapters and take a look at the emergence of life, the phenomenon that most closely involves us in the great symphony of the universe.

If dawn is the beginning of the day, the first instant in which something new "happens" is the beginning of a new era. I like to think about the moment when a hominid, our ancestor, eyes lifted to the sky, first asked the question: "Where do we come from?" Together with the question about our ultimate destiny, both as individuals and as a species, this is the question we've been asking ever since that far distant past, which we struggle even to imagine but that, for some strange reason, feels close to us. In fact, although scientists and philosophers have tried to provide a variety of answers in many different ways, from a strictly scientific perspective, we haven't made great strides. And this is in spite of the immense technological progress recorded over the last century.

So, what do we know about the origins of life? Let's start with what is commonly accepted. The phenomenon of life, as we know it on Earth, is due to an *abiogenesis* mechanism in which nonliving matter gradually organized into complex molecular structures capable of carrying out self-replication, self-assembly, and autocatalysis processes that resulted in the emergence of cell membranes, a fundamental element in the construction of living structures. If this is the case, the question becomes: Is life on Earth the result of a process of local abiogenesis or was it transported to Earth on a comet or an asteroid?

The second hypothesis is called *panspermia* and assumes that life in the universe is abundant, that it tries to fertilize all celestial bodies, and that its development depends solely on the environmental conditions of the host planet.

We don't know the answer.

We have a number of clues that lead us in one direction or the other, but no definitive proof. For example, fossil evidence suggests that life may have been present, at the level of single-celled organisms, as early as 3.7 billion years ago, when the Earth was a young planet formed less than 800 million years previous. There are even indications, albeit controversial, of life dating back to 4.28 billion years ago, about 100 million years after the formation of the primordial oceans, environments that were protected enough to allow for the development of life. Analysis of the structure of living species' DNA also suggests notable similarities that point to the existence of a common ancestor from which the various species have gradually differentiated. Such indications give credence to the hypothesis that life on Earth developed very early and that, starting from the oceans, it rapidly colonized the entire planet and then evolved over the course of billions of years into the complex multicellular organisms typical of the animal and plant kingdoms. On the other hand, the distance—in terms of complexity—between chemical and biological processes is such that we cannot yet delineate a credible abiotic process for the emergence of life. In 1952, the American biochemist Stanley Miller carried out an experiment: using an ampoule containing a mixture of liquids and gases, he tried to reproduce an environment identical to what is thought to have been the Earth's so-called *primordial soup*. Starting with only inorganic molecules subjected to electrical discharges, he proved that the primary amino acids, the organic building blocks of which proteins are composed, were produced. In his original experiment, Miller detected eleven of the twenty amino acids used in biological processes, but some recent reanalysis of his data has indicated traces of all twenty. This data is significant, but it only touches on the initial phase. The biochemical process that allows for the development from amino acids to actual cells is very long and it is not easy to evaluate, in terms of probability, whether this actually occurred on our planet over the course of a few hundred million years. To create living structures that are able to self-replicate would require an impressive, almost trillion-fold increase in amino acid complexity.

Let's take a look at how we arrive at this estimate. Amino acids are typically formed of around twenty atoms and have an average atomic weight of 110 mass units. A first factor of one thousand takes us from amino acids to proteins. Proteins can include up to tens of thousands of amino acids: in terms of atoms, that's around a half a million. The complexity of viruses is comparable to that of proteins: the smallest virus is made up of about 180,000 atoms. However, a virus cannot yet be considered a living organism in that it needs a *host*, which is to say, a cell, in order to reproduce. A second factor of one thousand takes us from viruses to bacteria, the smallest single-celled living organisms made up of hundreds of millions of atoms. In turn, a bacterium is much smaller than a cell in a higher organism. Here the leap in complexity is a factor of one million: a human cell is made up of around 100,000 billion atoms. Unquestionably, these biochemical processes, once activated, are stable and robust, despite their extreme complexity. An excellent example is chlorophyll, omnipresent in plant organisms. It is based on a series of biochemical reactions so intricate that scientists have been studying them for more than a century without fully understanding them. Although research in this branch of chemistry has led to at least thirteen Nobel Prizes, there are aspects that still need to be clarified. This is to say that today, starting from the basic principles, we are not able to design a mechanism analogous to chlorophyll synthesis; yet nature itself, a long time ago, developed this process along with its innumerable adaptive variants. This doesn't mean much in and of itself; it only serves to describe and remind us of our current level of ignorance. At the same time, however, it is a fact that makes us think. It even raises a doubt: Has nature reached the levels of complexity characteristic of life by pure chance or is it an inevitable path, as yet invisible to us, that leads from the simplicity of molecular chemistry to the complexity of the cell in a reasonably short time?

Steven Wolfram, a theoretical physicist known for having developed Mathematica, a type of powerful formal logical-mathematical analysis software widely used around the globe, has been studying algorithms known as cellular automata for a long time. These are relatively simple iterative algorithms in which the state of one cell depends on the states of the adjacent cells. It is something like a more complex version of the Game of Life invented by mathematician John Conway in the late 1960s. Wolfram's research shows how the evolution of certain algorithms leads to the emergence of stable local structures, that is, conditions in which the complexity of the structures formed by cellular automata is maintained for a long series of iterations before proceeding, according to random mechanisms, toward another stable local structure, often increasing in complexity. The aspect on which we should focus is that a similar process is also involved in the evolution of living species. With current genomic sequencing technologies it has been possible to prove how the DNA evolution of single-celled organisms from the same culture progresses through islands of genetic stability, and not necessarily through the same sequence of intermediate steps. These stable local structures are a sort of Japanese bridge, a series of stones solidly embedded in a watercourse, which allow you to pass, perhaps after multiple trials and errors, from one side of the stream to the other. So, the concept of stable local structures is a reasonable way to describe the process that could lead from inorganic chemistry to the complex biochemistry of the cell. For example, the cell's metabolism is based on a macro-component like RNA, a macromolecule similar to DNA but able to reproduce, which could represent one of the intermediate points in the direction of the cell's development. In this sense, based on our current knowledge, the two theories of abiogenesis and panspermia are both considered possible. The emergence of life in the universe would just be a matter of probability and therefore of time. The open question concerns the number of attempts and the time it takes to reach an organism complex enough that it is able to reproduce. The time and possibilities available on a single planet are immeasurably fewer than those available on the billions of planets that exist within a galaxy. So, if life tends to develop "rapidly," there would be no need for the panspermia theory; life in the universe would develop on its own, wherever the appropriate conditions exist.

To determine what "rapidly" means, we can use the time it took life to appear on Earth as a benchmark: a few hundred million years. Otherwise, if it took a much longer time for life to emerge, we would have to assume that once life appeared in some part of the universe it could have moved from one planet to another, from one solar system to another. But how could life move from one place to another, crossing the vast spaces that separate the planets and the stars?

15

INTERSTELLAR MIGRATIONS

By the time we realized that there was an extrasolar intruder, 'Oumuamua, named after the Hawaiian word for "scout," had already passed its closest point to the Sun and was leaving, as fast and stealthily as it had arrived. We are talking about the first sighting, in 2017, of an asteroid from another area of the galaxy, a messenger from distant worlds. What do we know about this dark, probably cigar-shaped shard, which visited our solar system with a trajectory and velocity that allowed it to leave so quickly?

Very little. We know that it was not made of ice, so it must be of the rocky type. It did not ignite like a comet as it approached the Sun. We know that it does not emit electromagnetic radiation. The most powerful radio telescopes have found no trace of it. Its orbit is gravitational, determined by the attraction of the Sun; a small, non-inertial component can be explained by the effect of the pressure of the radiation in our star's vicinity. We know that its speed, before entering the solar system, was compatible with the characteristic speeds of celestial bodies in the region of the Milky Way, of which our solar system is part. This allows us to exclude the idea that it comes from one of the dozen stars closest to us, as its velocity would have been too high. However, we have identified four more distant stars near which it could have passed in the last million years, with a velocity low enough that it could have originated in one of these star systems. So, we don't know exactly where it comes from, if it has already been in our solar system, how many other systems it has visited, or its composition. According to one hypothesis, it could be a fragment of an exoplanet destroyed by tidal effects. In this case it would be an object much rarer than main belt asteroids or objects from the Oort cloud, which formed directly from the original nebula. What is certain is that, on timescales of the order of millions or tens of millions of years, fragments like 'Oumuamua can bring different star systems into contact. One estimate even predicts that 10,000 extrasolar asteroids cross Neptune's orbit on a daily basis.

It would be interesting to be able to explore one to see what it was made of. This type of asteroid would seem to be the kind of vector suitable for transporting life, in hibernating form, from one part of the galaxy to another. While a space mission of this kind would be difficult because of the speed at which these fragments are moving, it wouldn't be impossible, considering that in the future our observational capacity will improve considerably, allowing us to identify these bodies sooner than we were able to identify 'Oumuamua. Another idea has to do with the possibility that some of these extrasolar objects have become trapped in our solar system after having lost some of their energy in a close encounter with Jupiter; a few candidates have already been identified. This approach would make an exploratory mission much easier to accomplish.

However, even the planets in our own solar system are in communication and exchanging material at a fairly high rate. Not everyone knows that we have about ten rock samples from Mars here on Earth, even though there has not yet been a mission that brought back material from that planet. The meteorite bombardment on Mars results in fragments that, given its thin atmosphere, can be projected into space. Some of them can reach the Earth, penetrate our atmosphere, and fall like normal meteorites. By comparing the isotopic composition of various meteorites with those measured on Mars during NASA's robotic missions to the planet, we are able to identify and distinguish Martian meteorites from all the others.

Finally, we should remember that it takes the solar system about 220 million years to revolve around the center of the galaxy. Since it formed 4.5 billion years ago, it has made the full circuit about twenty times. This means that, in the timescale in which life emerged on Earth, the newborn solar system made at least three complete circuits, coming into contact with fragments from distant star systems.

In 2019 I participated in a Breakthrough Discuss conference in Berkeley on "Migration of Life in the Universe." I was puzzled by the conference theme: we know almost nothing about life in the universe, I thought, so how we could talk about migration of life? But recalling the observation of 'Oumuamua, I did participate and I am glad I did. I was surprised by the scientific quality of the talks and by the extreme fascination of the topic. Life probably doesn't need massive, rocky starships to move from one planetary system to another. Considering the minuscule size of bacteria, the smallest living organisms we know, or even viruses, which can live and reproduce inside bacteria, we can also imagine other mechanisms suitable for this kind of transport. Microscopic ice crystals and dust, for example, containing bacteria and spores capable of withstanding the conditions in space, can spread into space from areas of a planet's upper atmosphere. When the dimensions become microscopic, the relationship between gravitational force, which is dependent on mass, and the thrust due to stellar radiation, which is dependent on surface area, tips the balance in favor of the latter. It is as if a planet were leaving a trail of perfume behind it. Planetary dust containing hibernating life can be pushed by radiation until it reaches high velocities and moves beyond a given star system, spreading to other systems or nebulae, where it can find suitable conditions to reproduce and evolve. We are used to thinking of space as vast and mostly empty, completely unsuitable for life. Perhaps we should change our minds. Space is less empty than we might think. In reality, the different parts of the galaxy communicate by exchanging material on timescales comparable to those of the appearance of life on our planet.

But how possible is it for life to survive in space? Well, even here, nature surprises us. In fact, we know of various living species that can endure extremely hostile conditions such as those in space: a nearly perfect vacuum, extreme temperatures, and ionizing radiation. Different kinds of lichens, bacteria, and spores are able to survive, losing all of their water and entering into a condition of total inactivity—which can last for extremely long periods—from which they can emerge, once they find themselves in a humid atmosphere again. Tests of this kind have been done on the International Space Station and in various laboratories. Even plankton, made of more complex organisms, shows a capacity to resist these prohibitive conditions.

A truly extraordinary case is that of the tardigrades. These very common micro-animals are about a half a millimeter long and live in water. They have eight legs, a mouth and a digestive system, as well as a simple nerve and brain structure. They are also able to sexually reproduce. They exist in nature in thousands of different versions and have a metabolism with unique characteristics. In order to withstand prolonged drought conditions, their bodies can achieve complete dehydration, losing around 90 percent of their water and curling up into a tiny, barrel-shaped structure. In other words, it's as if they freeze-dry themselves. Once this process is complete, their metabolism becomes 10,000 times slower. The most amazing thing is that they can stay in this state for decades, only to wake up again within twenty or thirty minutes once exposed to moisture. But there's more. When in a dehydrated state, they can withstand the vacuum of space as well as pressures higher than normal atmospheric pressures, temperatures near absolute zero or temperatures up to 150°C. Their radiation tolerance threshold is hundreds of times higher than what would be deadly for humans. The secret of their ability to harden is due to a sugar, *trehalose*, which is also widely used in the food industry. When dried, this sugar replaces the water molecules in the cells, leaving the animal in a kind of vitrified state.

In addition, the tardigrade's DNA is protected by a protein that reduces radiation damage. Is this information enough to make us assume that these micro-animals come from space? I would say no. Their unusual metabolism is more likely the result of evolutionary adaptation that happened on our planet. In fact, tardigrades are among the very few living beings that have emerged unscathed from all five extinction events that have occurred on Earth. That is why they are the best candidates for a long journey into space aboard a meteorite or a comet. Recently, tardigrades have achieved a bit of media notoriety resulting from the Beresheet mission, a private probe launched by Israel, that crashed on the Moon in early April of 2019. The probe was carrying a colony of these micro-animals, in their dehydrated state. Given their microscopic size, it is likely that they survived the crash and will remain inactive for a long time to come, ready to be reawakened from their hibernation. By replacing the Israeli probe with an asteroid, we have a textbook example of how life might have arrived on Earth.

Or how life could have migrated from Earth to other planets in our galaxy.

So, the problem of the origin of life remains open, even if, step by step, we are making progress toward a solution. In the last decade, increasingly powerful calculation instruments have allowed us to reproduce, starting from the first principles of quantum mechanics, the formation of increasingly large and complex molecular systems, now made up of thousands of atoms. The field of computational biology is growing at a formidable rate; it is now only a matter of computing power.

At the same time we have dramatically developed our ability to decode and manipulate DNA, up to the creation of the first simplified genomic structures, derived from living organisms and able to reproduce. We are now talking about synthetic life, built around human-designed DNA, a field with huge development prospects.

Therefore, it is likely that the creation of the complex molecular structures needed for life or the confirmation of the existence of islands of genomic stability in the evolution of viral and bacterial species are objectives that, in future, will be within our reach. At that point, we will have another tool for understanding how life on Earth developed. Who knows? Perhaps we will discover that aliens are particular biological life forms that have lived with us since the beginning of time; and we were looking for them on Mars or below the icy surface of Jupiter and Saturn's moons!

16

OTHER SUNS, OTHER WORLDS

While our solar system was forming in this corner of the universe, there is no reason to think that the rest remained idle. The idea of extrasolar planets, as we know, is certainly nothing new. The problem is that the very complexity of the formation process of planets and stars means calculating the probability that a given star is accompanied by one or more planets is still unreliable. Giordano Bruno, a brilliant sixteenth-century visionary thinker, went far beyond Copernicus in hypothesizing the existence of other worlds, each one of which could presume itself to be the center of everything, be inhabited, and be better than our own. In his 1584 work *On the Infinite Universe and Worlds* he writes: "[It] is one, the heaven, the immensity of embosoming space, the universal envelope, the ethereal region through which the whole hath course and motion. Innumerable celestial bodies, stars, globes, suns and earths may be sensibly perceived therein by us and an infinite number of them may be inferred by our own reason. The universe, immense and infinite, is the complex of this [vast] space and of all the bodies contained therein."

Isaac Newton himself, in formulating the law of universal gravitation, spoke of the possible existence of other planetary systems. However, we have to wait until the early 1990s before we have evidence of the first exoplanet. In 1992 two radio astronomers, Aleksander Wolszczan and Dale Frail, identified two planets orbiting around a pulsar, a rapidly rotating neutron star, probably formed by the explosion of a supernova, in an environment very different from the one in which our solar system formed. In this case, it can be hypothesized that these are solid core frag- ments of gas giants that survived the explosion of the supernova and entered the pulsar's orbit. Because of this, the first real observation of an exoplanet is considered to be the one announced in 1995 by two Swiss astronomers, Michel Mayor and Didier Queloz, who detected a planet around the star 51 Pegasi. This star, like the Sun, is a main-sequence, G-type star, so we can expect the same mechanisms that led to the forma- tion of our solar system to be at work here.

So, five centuries later, Giordano Bruno's visionary insight, which together with a number of other ideas that went against the thinking of the time and led him to be burned at the stake, has become scientific evidence. Not only that; today, the speed at which discoveries are being made in this area is growing dramatically. The parameters that character- ize the properties of the initial nebula (chemical composition and tem- perature), the central star (size, brightness, and stability), as well as the gravitational dynamics between the planetesimals and forming planets, are so many and connected in such a complex way that it makes us think of an incredible number of possible solutions. And this is exactly what we observe when discovering one planetary system after another. As of 2019, we knew of the existence of more than 4,000 exoplanets, and 3,000 other possibilities are under investigation. The rate at which exoplanets are discovered doubles every few years and the variability of their char- acteristics turns out to be substantial. In 2020 the first extragalactic exo- planet was observed as well as an Earth-mass planet unbounded to any star, running around in our galaxy. Space missions dedicated to exoplan- ets follow one after the other: the European Space Agency (ESA) alone is developing three.

While there is no reason to believe that the Earth is a "typical" planet, among thousands of observed cases we have evidence of planets with surprising characteristics. Let's focus on the planets that are rocky like our own; a question that we can ask is if the climatic conditions could sup- port the presence of liquid water on the surface. Be aware that this does not mean that there is water, just that the temperature exceeds zero, at least for a certain period of the year and in a certain area of the planet.

If the star's characteristics are known, it is possible to define a distance interval within which those conditions exist.

This range of orbits, which are compatible with water in the liquid state, is known as the *circumstellar habitable zone* (CHZ), or simply the *habitable zone*. In the case of the solar system, the Earth is well placed at the center of the habitable zone, but Mars and Venus are also within the boundaries. However, we know that there is no liquid water on the surface of either of those planets today.

Why not? In the case of Mars, about 3 billion years ago its atmosphere was already almost completely gone, perhaps due the disappearance of the magnetic field that protected it from the bombardment of solar wind and cosmic rays. We know that there was once a great deal of water on Mars, but now on the surface there is only ice covered with a layer of soil, although deep underground lakes of liquid water have recently been discovered by an Italian team using Mars Express radio imaging data.

As for Venus, we are seeing a textbook case of the *greenhouse effect* on a grand scale, which accounts for its extremely high surface temperature. Venus is, in many ways, similar to Earth, but it has a very dense atmosphere made up of carbon dioxide. The surface pressure is more than ninety times higher than what is found on Earth. The cause lies precisely in the very high temperature, averaging 480°C, which causes the planet to be trapped in something like a giant pressure cooker. We cannot exclude the possibility that Venus once held large quantities of water, but in the current conditions it would evaporate instantly.

It is clear that falling within the *habitability range* is not sufficient to identify planets capable of supporting life as we know it on Earth; it is only a precondition. The evolutionary history of the planet, in particular, the presence of mechanisms that protect it from its star's radiation, such as the presence of a magnetic field, and its composition, are equally crucial. If we think about Mars, it has been the red desert we know for most of its existence, but in the first one or two billion years, it was full of water. So, the probability that an exoplanet is habitable and/or inhabited depends on time and therefore its history, not only its physical properties.

It is evident enough that as the number of exoplanets we observe increases, the chances of finding something that looks like Earth also increases. As we know, this excites the public's imagination. Every now and then there are newspaper headlines that reveal the discovery of a planet the same as the Earth in every respect. Is this really the case? I would say not. In reality these exoplanets are about the same size as the Earth and located at similar distances from stars that are comparable to the Sun, but to say that there could be life on these planets is overstating things. It is reasonable to expect that, sometime in the future, we will have more detailed information on each exoplanet, especially when it comes to the atmosphere.

In the case of stars whose exoplanets rotate on a plane on which the Earth–star line of observation lies, the most effective detection technique is one based on the slight decrease in the star's brightness when the planet passes in front of it. It is a clear signal, easily distinguishable with our current technological instruments. As it crosses in front of the star, even the thin layer of a planet's atmosphere filters the light passing through it. Ongoing experiments now aim to measure the effect of this filtering and deduce the atmospheric composition. In order to do this, a noticeable improvement in the sensitivity of our instruments is needed; however, it is reasonable to think that we will achieve this soon. The first evidence of water vapor in the atmosphere of a distant exoplanet was recently published. Based on indirect reasoning, we are already able to obtain indications about the characteristics of some exoplanets. Complex planetary systems have been discovered recently, such as the seven planets of Trappist-1, a red dwarf located about 39.5 light-years away in the constellation Aquarius. It is significantly smaller than the Sun, with about 8 percent of its mass, a surface temperature that is less than half that of the Sun, and it is slightly larger than Jupiter. Most of Trappist-1's planets fall within the habitable zone, and one in particular has an Earth Similarity Index (ESI) of 0.90, among the highest yet found (to be clear, the maximum ESI is 1, that of Earth itself).

Considering that Mars has an ESI of 0.80, in the case of Trappist-1e, we are talking about a relatively welcoming planet, which probably features a substantial amount of liquid water. So everything's okay, right? Not exactly. The planets of this small extrasolar system are probably in synchronous rotation, always showing the same face to the dwarf star. In short, there would be significant differences in temperature between the day side and the night side, causing violent and continuous winds.

The more temperate zone, suitable for any possible development of life, would be the twilight zone, the one that marks the passage between day and night. Another element to consider is the activity of red dwarfs, often subject to much more dramatic flares than are G-class stars such as the Sun. This would jeopardize the presence of planetary atmospheres on satellites of Trappist-1. That said, thanks to observations from Earth and from space, particularly those made with NASA's Kepler satellite, we have discovered thousands of exoplanets. In fact, astronomers have gone even further. They have decided to compile a kind of "bestiary" in which the planets with the strangest, even the most bizarre, characteristics are collected.

A few examples? There is an exoplanet that seems to be pink and another with a perfectly black surface. Some have orbits too close to their star and, consequently, are evaporating. Others have startlingly high surface temperatures. In some cases rocky planets have been observed that, due to tidal effects, always have the same side turned toward their star, like the Moon relative to the Earth. On one side the surface temperature exceeds that of molten lead, while on the other it is the freezing cold of deep space. I'll let you imagine what happens in the twilight zone (also called the *terminator*), where drops of lead and tin rain from thick metallic clouds. Then there are planets that are located in the habitable zone but which have such long, elliptical orbits that at the *aphelion*, the point furthest from the star, all of the water is frozen, while at the *perihelion*, the point closest to the star, it evaporates into the atmosphere. On Earth, it would be as if in summer all the seas, rivers, and lakes evaporated and became a thick blanket of clouds and then, in autumn, torrential rains returned all of the water to the planet's surface for a very long winter in which it all turned to ice.

The composition of the original nebula might also be very different from that which gave rise to our solar system. Since carbon is the fourth most abundant element in the universe, after hydrogen, helium, and oxygen, it is plausible to imagine nebulae in which carbon is dominant over oxygen. In that case, the protoplanetary disk would form solid compounds based on silicon and titanium carbides rather than silicon and oxygen. Planets composed of this material could still have an iron-rich core like the Earth's, but at the same time a graphite surface that, if the pressure was high enough, would be covered with a layer of diamond. Then, if there were volcanic activity the eruptions would form mountains of diamonds and silicon carbide.

With a certain frequency, stars form or bond in pairs or even triplets, but this does not prevent planetary formation. Numerous exoplanets orbit around binary stars, a condition in which the concept of sunrise and sunset becomes considerably more complicated. In particular, the Earth's closest star, Proxima Centauri, is part of a ternary star system: Proxima Centauri is a small red dwarf while Alpha Centauri A and B are two dwarf stars, yellow and orange, respectively. The discovery of a rocky planet orbiting Proxima Centauri was announced in 2012. This is all the more surprising since the fact that even the star closest to the Earth has at least one planet would suggest that stars with planets are the rule rather than the exception. One thing is certain: the study of exoplanets also helps us to better understand the mechanism of planetary migration. Let's think about the planets known as *hot Jupiters*, gas giants orbiting close to their star, inside the frost line, a position in which they could not have formed. This forces us to hypothesize a process of migration from the outside toward the inside, connected to the dynamics of the planet's growth. In other cases, we have seen rocky planets migrate to the other side of the frost line, which we talked about in chapter 12. In this case, the migratory dynamics are in the opposite direction. In the exoplanet zoo there really is something for everyone. Certainly, the future will hold other surprises; thanks to increasingly powerful and sophisticated observational tools, exoplanetology is one of astrophysics' fastest developing fields.

Moreover, the complexity of planetary system formation is such that observational studies are essential for validating theories and models. For millennia we could only reflect on our own solar system and establish hypotheses about its formation. However, today we know that many, if not most, of the hundreds of billions of stars in the hundreds of billions of galaxies that surround us are accompanied by planets. Soon we will be able to count on an endless supply of data that we can use to compare predictions and theoretical models, moving ahead in our understanding of planetary formation.

17

WHERE ARE THEY?

Chicago, summer of 1950. Legend has it that Enrico Fermi, while going to lunch with Edward Teller, Herbert York, and Emil Konopinski, not only colleagues but also friends, was chatting about the recent wave of UFO sightings reported in the American press. During lunch the famous Italian physicist who, as you may well imagine, was not very convinced by such generic and optimistic reports, suddenly returned to the subject and asked the others this provocative question: "Where are they?" The phrase has secured its place in history as the Fermi Paradox. He tried to calculate the probability that our planet had been visited by other intelligent beings, and concluded that the encounter should have taken place several times already. Hence the paradox that bears his name: If the odds are in fact so high, why have we not yet been able to verify the existence of extraterrestrials?

Fermi died of cancer a few years later, in 1954, but his question continued to intrigue scientists. A decade later, Frank Drake wrote a simple equation that sought to formalize the calculation of the number of intelligent species in our galaxy. It's interesting to take a look at this equation, especially in order to understand its limits. The formula aims to determine the likelihood that such a species developed and made a concerted effort to communicate its presence. By multiplying this by the number of habitable stars and planets in our galaxy, we can calculate the number of

civilizations that could have visited the Earth since its inception. Therefore, the equation is the product of a series of numbers and probabilities:

$N = R^* \times fp \times nh \times fl \times fi \times fc \times L$

The parameters of this equation are astrophysical, biological, and technological, respectively.

The first three can be calculated with relative precision: R^*, the rate of
formation of stars in our galaxy; fp, the fraction of stars that have planets;
nh, the fraction of stars that can potentially host life. According to current estimates, the first two parameters are on the order of 1 and the third
on the order of 10 percent. In Fermi's time, the uncertainty around the
average number of planets per star was much greater than it is today and
therefore the values estimated were very low.

The second two parameters are biophysical, and here the uncertainty
increases drastically: fl, the fraction of planets on which life has actually
emerged at some point in their existence, and fi, the fraction of planets on which intelligent life and a sufficiently evolved civilization has
developed. Those who support the abiotic hypothesis think that the first
parameter is a number potentially not far from 1; life tends to emerge from
inanimate chaos on its own. The second parameter is utterly uncertain,
also because the definition of evolution or intelligence is rather vague.
On our planet, for example, other existing species demonstrate behaviors
corresponding to some level of intellectual development; just think of
dolphins, elephants, or certain types of great apes. For some years now
there has even been discussion about whether plants, an example of living systems without a brain, have developed their own type of individual
or group intelligence, in their interactions with the environment.

The last two parameters are sociological/technological ones: fc the
fraction of civilizations that reach the technological level at which signals indicating their existence can be sent into space (using electromagnetic waves, neutrinos, interstellar probes), and finally L, the length of
time these civilizations have been able to keep up this kind of transmis-
sion. Here uncertainty reigns supreme. We have only one example of this
ability and it is, have a guess: our own species. On the basis of a single
example we can reason as much as we would like, but it would be difficult
to achieve a worthwhile result. However, we can consider some factors. If

life could develop over the course of 100 million years, our species came into being much later than that, at least 4 billion years later. From an evolutionary point of view, the decisive steps have occurred *only* in the last 2–3 million years. Not to mention advanced technologies; radio waves, the only way that we know of to easily communicate over interstellar distances, were discovered just a century ago. As for the future of human society, we have begun to have serious doubts that very slow evolutionary development and certainly not rapid development of civilizations can truly be compatible with the lightning-fast and exponential technological advancement we are currently experiencing and with equally rapid population growth. It's enough to think of climate change and reaching the limits of human sustainability on this planet, aspects that I have already pointed out and which are or, it is hoped, should be, not surprisingly, at the center of the political and cultural debate around the globe.

As a result of these uncertainties, the actual estimates of the number of advanced civilizations, in addition to our own, range from numbers close to zero to those exceeding one million. With all due respect to Fermi, "Where are they?" remains an open question without a useful answer. As for Drake, his equation is essentially a tool for keeping track of our ignorance about this issue.

Writers and sci-fi film directors can rest easy: for some time to come their imagination will have to stand in for our knowledge.

18

THE DARK SIDE OF THE COSMOS

Let's get back to the evolution of the universe, this time dealing with its dark side, which, as we will see, turns out to be the dominant one. In our discussion on the origins and evolution of the universe, we should avoid falling into the classic error of the person looking under the lamp for the keys even if they have been lost elsewhere, just because that is the only place with enough light to see. The similarity, although it might seem strange to some, is based on more than one element. The fact is that the first stars and supernovas constitute the "visible" part of the unceasing work of gravity at every scale. But the cosmos contains much more than that. There are the "invisible" parts, which are just as if not more important, and they need to be considered if we want to understand how the universe has evolved up until now. First, we saw that the metric of the universe—that is, the distance between two points in space-time—has never stopped expanding and will continue to expand for a long time to come, some argue forever.

On the other hand, at local dimensions, that is, galaxies and groups of galaxies, it is gravity that regulates the formation of structures at every possible scale. Toward the larger end of the scale: stars, galaxies, groups of galaxies, but also toward the smaller end of the scale: planets, satellites, neutron stars, black holes. Some of them are easily visible because they are bright; others are observed without our being able to identify their position, like some types of neutron stars, called pulsars, which emit pulsating electromagnetic radiation but are difficult to locate with optics or X-rays. There are still others, like planets revolving around other stars, visible only thanks to sophisticated observational techniques, which have been available only for a few decades.

Other structures are just not seen, they can be observed only through the effects induced on bodies orbiting in their vicinity; here we are talking about black holes, clusters of nonluminous matter, or clouds of cold gas and dust. When it comes to the filaments of matter that connect different galaxies or dark matter structures present at all scales, we can identify them only through computer simulations for now. While we might be surprised to find that most of the universe is dark, and therefore invisible, just think about looking at a plane flying above the Earth's surface at night: we can recognize only the parts of the plane that are illuminated by lights. Looking down from the airplane, cities and highways lined with streetlights stand out vividly while to our eyes peripheral zones, often covering vast areas, which separate one city from another, appear devoid of any point of reference because they are completely in the dark. So, we must wait for daylight to come in order to understand what is contained in the space that separates two cities. But we cannot do the same with the universe. There is no impending daylight.

Take *dark matter* and *dark energy*, for example. Observing the universe has allowed us to understand that there are two invisible components that determine the dynamics of stars and galaxies at both large and small scale. As previously mentioned, on a large scale, the rate of metric expansion accelerates, probably due to the activation of a further component of vacuum energy that causes, as in the case of inflation at the beginning of time, a repulsion that further expands the metric. In this case we are talking about a much weaker effect that becomes observable only when the distance between two objects is "cosmological," that is, comparable with the visible size of the universe. This component is called dark energy; it was identified in the last twenty years, and, in 2011, it earned Saul Perlmutter, Brian Schmidt, and Adam Riess the Nobel Prize in physics. In the matter and energy balance of the universe, dark energy weighs fifteen times more than atoms, nuclei, electrons, light, neutrinos, and antimatter put together.

At a small scale, however, there is systematic evidence that the movement of celestial bodies (stars in galaxies, galaxies in galaxy clusters) is influenced by the presence of an invisible mass, called *dark matter*, which is gravitationally active. This phenomenon is very visible in galaxies.

We know that gravitational attraction is due to the mass contained in the volume that is in the sphere below the orbiting object. If we analyze a globular-type galaxy—which means a galaxy in which the stars are arranged in a spherical way—it is reasonable to think that once the typical visible radius of the galaxy has been exceeded, the attractive mass reaches its highest possible value. Consequently, stars orbiting peripherally with respect to the galaxy should all be affected by the same attractive mass, regardless of the distance from the center of the galaxy. Instead, we observe that the peripheral stars move as if the mass increases more with distance, an evident sign of the presence of mass that does not emit light but that is as gravitationally active as the visible mass.

Evidence of this anomaly has been known for nearly a century: it was reported in 1933 by Fritz Zwicky, a Swiss astronomer at the California Institute of Technology, and has since been observed in thousands of galaxies, in particular through the seminal work of Vera Rubin at Georgetown University in the early 1950s. It has also been noted at other scales, such as the scale of the interaction between two different galaxies, or that of galaxy clusters. The quantity of dark matter necessary to explain these effects is about six times greater than that of ordinary matter.

Obviously, the fact that all of the physics research on the properties of matter concerns a type of matter that, in astrophysical terms, turns out to be a small fraction of the total is worth highlighting. Scientists have been trying to find a solution to this problem for a long time, mostly through careful study of the cosmos, the only place from which evidence of dark matter is available. Many hypotheses have been formulated: the presence of an anomalous number of giant planets, for example, Jupiters, or of other concentrations of ordinary cold matter, of objects invisible to telescopes but effective from a gravitational perspective. However, careful observation of the skies has ended up limiting the number of these objects, and this hypothesis was eliminated. The same reasoning was offered for black holes, the number of which is difficult to estimate. Since the gravitational waves propagated by the collision of black holes have been observed, the density of these objects has been more accurately determined—at least within a certain range of mass values—but it does not seem sufficient to explain the necessary quantity of dark matter. Another solution has been proposed that would probably have both Newton and Einstein rolling in their graves: a theory of Modified Newtonian Dynamics (MOND), a correction of gravitational force at distances on the order of the size of galaxies, in order to adapt it to match observational data. The more precise the observations, however, the more difficult it is to argue the validity of this alternative theory.

According to yet another hypothesis, the explanation lies in the existence of elementary particles with mass and therefore gravitationally interacting, but weakly or not sensitive to the other three fundamental interactions: electromagnetic, as well as weak and strong.

Various indications point in this direction. First of all, astrophysical-type measurements are compatible with a model of dark matter due to elementary particles with large mass produced during the Big Bang but never directly detected. These particles would be everywhere in the universe, but they would also be concentrated in the regions where galaxies formed: a diffuse halo of these particles would extend around all galaxies, playing an important role in determining their shapes (spherical, elliptical, with arms, etc.).

Second, the Standard Model of Fundamental Interactions (which we will discuss in the next chapters) describes the properties of the fundamental forces and particles that correspond to less than 5 percent of existing matter. This model can quite naturally be extended to include the new type of particles corresponding to dark matter. The idea of this extension arises from the fact that the analysis of the properties of the interactions between elementary particles is easy when we introduce particular properties of symmetry that are respected by the interactions themselves. A very simple example is symmetry by reversal of electric charge: interactions between charged particles are invariant if all the signs of the electric charges are reversed. In the case of dark matter, the hypothesis is that there is a supersymmetry capable of describing all three of the fundamental interactions. In this context, the theory predicts the existence of other elementary particles, *supersymmetrically* corresponding to the known ones, including one neutral (the lightest), which would be stable and could be the source of the mysterious dark matter. This particle is generically called a weakly interacting massive particle (WIMP). The search for evidence in favor of supersymmetry, long pursued at Geneva's Large Hadron Collider (LHC), has not yet met with success. We have no indications of the mass of this particle; depending on the theory proposed, it could range from very small values—comparable to the mass of neutrinos—up to thousands of times larger than the mass of a proton. Since there is an astrophysical indication of what density of dark matter is needed to complete the mass–energy balance of the universe, the larger the mass of this particle, the less of it there is around.

Dark matter has been the subject of ongoing systematic research for many years now; we are searching for it, literally, with every possible technique.

Using accelerators like the LHC, we are seeking to harness the potent energy available in the incident beams to produce WIMP particle/antiparticle pairs. The presence of these pairs would be observable from the experiments constructed around the interaction zones. But we are also looking for traces of WIMPs in underground experiments conducted at laboratories like the underground laboratory at Gran Sasso, operated by Italy's National Institute for Nuclear Physics (Istituto Nazionale di Fisica Nucleare, or INFN), where special instruments are capable of detecting the very rare collisions in which one of these particles interacts with normal matter, leaving an energetic deposit (direct detection). Finally, we look for traces of them in space experiments, studying the rare components of cosmic rays with precision, looking for possible contributions due to the annihilation of WIMP pairs in the cosmos (indirect detection). As of yet, all this effort has yielded few results. No convincing and definitive signs of the existence of dark matter have been recorded with any of the three methods just described. However, some interesting effects have been found that require an explanation, particularly in the case of indirect detection, which we will discuss later.

The fact remains that, today, the dark part of the universe dominates the visible one. All of the galaxies are surrounded by a halo of invisible matter, which has determined their shape and size and influences their evolution. The types of matter and energy that dominate the dynamics and structuring of the cosmos are different from those of which we are made. We know that we don't know, as we said at the beginning of this book. We move ever further from the idea of our centrality in the cosmos; the matter we are made of doesn't even make up part of the main component of the universe.

19

THE ORGANIZATION OF THE UNIVERSE

While space continues to expand on a cosmic scale, we have seen that the organization of matter on a local level proceeds relentlessly, at all scales, under the discrete and omnipresent direction of gravity. Remember that any mass, whether visible or invisible, is a source of gravitational force and is subject to its effects. But we must not forget that matter is also affected by the other three fundamental forces existing in nature: electromagnetic, weak, and strong. The effects of these forces may be easier or more difficult to observe, but they are of critical importance in the structuring of matter and in the evolution of the cosmos. Unlike gravity, the other forces mainly operate at the microscopic level, each with unique and extraordinary characteristics, which researchers have come to understand only over the course of the past 150 years.

Electromagnetic interactions are long-range forces, just like gravitational force. But these interactions are vastly more intense, about 10^{36} times more than gravity at the proton scale. Why is it that they don't dominate the structure of the universe instead of the much weaker gravitational forces? The reason is this: the intensity of the electric field is counterbalanced by the fact that there are positive and negative electric charges that tend to neutralize each other as quickly as possible, while gravity has only one kind of charge, mass, which can only increase. So, the electric force tends to "cancel out" its effects by binding together opposing charges, for example, electrons and protons forming neutral atoms. Electromagnetic interactions are carried by photons, the quanta of light introduced, as we know, by Einstein. Thanks to the photons, and the energy they carry, it is easy to create or break bonds between opposite charges. Electromagnetic forces also exhibit a short-range component, even when neutral atoms are involved; they are called Van der Waals forces and are very important in surface interactions between materials.

Strong interactions (or forces) operate only at short range, at nuclear or subnuclear dimensions of around one millionth of a billionth of a meter—a dimension typical of nuclear physics and that used to be called a fermi after the physicist Enrico Fermi (it is now called a femtometer, equal to 10^{-15} m). These forces hold together quarks, the elementary particles that form protons and neutrons; they also hold the protons and neutrons together in atomic nuclei. Strong force has a striking feature: it plays hide-and-seek with the experimenter. Quarks interacting due to strong force are not detectable on their own, but only in pairs or groups of three. It's as if I had a sack of walnuts: trying to open it and take some walnuts out, I find myself with two sacks containing walnuts but never with a single walnut in hand. During the very early stages of the universe, matter was so hot and compressed that this phenomenon, called *confinement*, could not be seen in the sea of elementary particles constantly and furiously interacting. It was detected when the temperature of the initial plasma dropped below a certain value and the quarks bonded to form hadrons. The components of atomic nuclei, protons and neutrons, are examples of these particles. Atomic structure is therefore ensured by the simultaneous action of two fundamental forces, strong force in the nucleus and electromagnetic force in the electron–nucleus bond, in addition to a pinch of quantum mechanics. In fact, if the atoms behaved like tiny solar systems, no two identical hydrogen atoms would exist, and, after a while, the electrons would precipitate into the nuclei and attach themselves to the protons, which clearly doesn't happen.

Finally, there are weak interactions: these interactions also occur at very small distances and typically transform elementary particles into each other. They are fundamental in nuclear processes that occur at the center of stars, or in neutron and unstable isotope decay.

Going back to gravity, it exerts its dominion over neutral matter that can be accumulated in unimaginable quantities; as the mass gradually increases, the attractive force increases. Starting from the smallest agglomerations of dust, gradually increasing in mass and size, gravity leads to the formation of asteroids, planetesimals, rocky planets, and gas planets of various different dimensions. As the mass increases, so does the pressure exerted on the planetary core by the weight of its outer layers. In our discussion on the formation of stars, we have seen how, depending on the elemental composition of the planet, as the mass increases, the internal pressure leads to the activation of nuclear fusion reactions that increase the temperature and produce a flow of thermal energy toward the exterior. This flow is accompanied by a pressure that tends to counterbalance gravitational force. If the mass continues to grow, exceeding one-hundredth the mass of the Sun, the internal fusion mechanism takes over and the planet transforms, becoming a protostar and then a star. Jupiter is a gas giant with a mass that is a thousand times smaller than the Sun's and 318 times larger than the Earth's. If it were just fourteen times more massive, it would have become a star, a brown dwarf, or, if it were eighty times more massive, a red dwarf.

But what happens if we keep adding mass to a small star? How much can a star grow? Stars can have masses that are eighty to ninety times smaller than that of the Sun, as in the case of brown dwarfs, or up to 150 times larger, forming a hypergiant star like Eta Carinae. This is a range that extends by a factor of 10,000 and covers stars vastly different in brightness, stability, and size.

In every case, the internal energy source is nuclear fusion. The bigger the star, the higher the pressure, and the internal temperature and the fusion reactions are intense as well as energetic. The largest stars are thousands of times shorter-lived than the Sun as well as being unstable. In hypergiant stars the flow of energy coming from within is so powerful that it periodically expels the surface layers of the star. When the fuel at the star's core runs out, it undergoes gravitational collapse of a violence commensurate with its size. Inside stars, atoms are separated into atomic nuclei, mostly fully ionized, and electrons. The star is neutral as a whole, so even if in plasma form, the overall volume is maintained.

The average density of the Sun is just 40 percent greater than that of water. When the core collapses beneath the weight of the outer layers, an electron capture nuclear reaction occurs in which electrons and protons fuse to create a neutron and an electron neutrino. This transformation eliminates the electric repulsion between protons and electrons and the star's core implodes into a state formed only by neutrons, with a density 10,000 billion times greater than the Earth's. The rest of the star explodes in a giant fireworks show. A vast number of neutrinos are emitted in this transition, playing the role of photons in cooling the early stages of the neutron star's life. If the star is rotating, the angular momentum—the quantity that measures how quickly the mass of a star is rotating around an axis—is conserved in the implosion of the core; it's like a skater who increases their speed by pulling the arms in toward their body while pirouetting. The small neutron star spins incredibly fast, thousands of times per second. But gravity never ceases to amaze us. If the collapsing core exceeds three solar masses, the result is something even more extraordinary: a black hole.

20

BLACK HOLES

Gravity often finds its equilibrium in playing roles. Planets ceaselessly orbit around stars thanks to the balance between gravitational force and the acceleration needed to cause a body to deviate from a straight line. We have seen that stars are the result of the counterpoise between gravity and the energy produced by nuclear fusion, in such a way that if the gravitational effect increases, the energy released by the fusion increases. Neutron stars—the state of matter that happens when the pressure is so high that it makes electrons and protons collapse, forming neutrons— find their equilibrium in the neutrons' resistance to being compressed further, once they are in contact with each other and repulsive electromagnetic forces are no longer present. But by continuing to increase the mass, the gravitational force can grow disproportionately; the feeble force of gravity turns out to be inordinately greedy and dominates anything that tries to oppose it. It is in these instances that it shows its invincible power. In the end, even light has to pay up. A photon that leaves a celestial body that has mass loses some of its energy in overcoming the gravitational potential of the body that it is leaving (the energy of a photon is equivalent to a unit of mass, albeit small). At some point, the body's mass increases enough to achieve an attractive force that does not allow even one ray of light to leave the source. At that juncture, the body becomes perfectly black, while still maintaining its ability to attract from a gravitational point of view.

The idea of a *black star*, a body so massive that it cannot emit light, was already formulated in a scientific communication at the end of the eighteenth century by John Mitchell, an English cleric and astronomer, and at the same time but completely independently by the famous French mathematician Pierre Laplace. The idea was right but, in addition to having a number of limitations, it was not based on the same conceptual foundations that, more than a century later in 1915, allowed Einstein to work out his revolutionary general theory of relativity. In Einstein's theory we find the formulation of a complex, elegant equation, in which the relationships between the metric of space-time and the distribution of mass and energy contained in it are defined. There are ten nonlinear differential equations, which are reduced to six with an appropriate transformation. These equations are the generalization of the relationship between mass and the force of gravity described by Newton's law of universal gravitation. Dealing with such a complex set of equations is anything but easy. Each solution can correspond to the most diverse conditions of space-time, matter, and energy, solutions that are often surprising when compared with those we were used to with Newton: gravitational waves are an example, the existence of different types of black holes another (we will talk about these shortly).

In those same months, between 1915 and 1916, the mathematician and astronomer Karl Schwarzschild found a solution to Einstein's equations that paved the way for a structure characterized by a radius limit, beyond which nothing could return, irresistibly attracted to the center. However, it would take a long time, until the 1960s, and with the contributions of a number of scientists, including Arthur Eddington, Georges Lemaître, Subrahmanyan Chandrasekhar, John Kerr, Robert Oppenheimer, Evgeny Lifshitz, Roger Penrose, Stephen Hawking, and others, to find a formalization of the concept of black hole. In any case, in order to have direct, experimental confirmation of the existence of black holes in our galaxy and others, we have to look to very recent times.

Some indirect evidence has been available for decades, but black holes were initially associated only with very bright and extraordinarily energetic objects, incredibly active galactic nuclei cataloged as *quasars* (quasi-stellar objects), or Seyfert galaxies. The exceptional brightness of these "monsters" of the heavens is most probably due to violent reactions involving the material surrounding the black hole in the accretion disk that feeds it. In the case of the Milky Way, the effect of the central black hole is less violent and visible. Studying the orbit of the stars near our galactic nucleus has nevertheless made it possible to establish the presence of a massive black hole around which the entire galaxy is revolving.

The black hole represents gravity's triumph; at its center there is a space-time singularity, a point in which gravity is so intense that nothing can resist it. Matter and kinetic energy disappear and are transformed into energy accumulated in the curvature of space-time. Therefore, the increase in the mass of the black hole corresponds to the further increase in this type of energy. Of course, the issue is not that simple. As we've seen, it took fifty years to find some solutions to Einstein's equations that correspond to the black hole. Moreover, as we gradually approach sufficiently small sizes, as in the case of the Big Bang, Heisenberg's uncertainty principle comes into play and, as we've seen, complicates matters even further. The general theory of relativity is based on the special theory of relativity, but does not take quantum mechanics into consideration. So far, all of the theoretical efforts made to draft a gravitational theory that is compatible with quantum mechanics have been unsuccessful.

There is no limit to the mass of a black hole: being defined by an intensity threshold in the gravitational field above which the process of gravitational collapse no longer stops, black holes can be very small but also gigantic. For those that are the result of the collapse of a star, there is a limit of about two solar masses; below this limit collapse will form a neutron star and not a black hole. However, once created, black holes can grow unchecked. At the center of our galaxy there is a black hole with an estimated mass of 4.1 million solar masses; but there are also others, hypermassive, with a mass of more than 10 billion solar masses. Although it may seem to the general reader that this is just word play, we can certainly say the only characteristic limit to the mass of a black hole is the time it takes to absorb all the mass it contains from the outside. According to some calculations, the largest hypermassive black holes may not have had the time to grow so much over the course of the lifetime of our universe. Naturally, this would pose a consistency problem with the

Big Bang model. Black holes have extraordinary properties, externally as well as internally. While what happens outside is observable, what happens on the inside is completely shielded from observation. In this case we can only be guided by the formulas of general relativity.

Let's start with what the general public sees as perhaps the most amazing property: the extreme curvature of space-time means that, as you approach the black hole, time slows down more and more until it stops at the edge of what is known as the *event horizon*, the area that delimits the volume from which light can no longer escape. From the point of view of an external observer, an object falling toward a black hole takes an infinite amount of time to reach the horizon. For the observer situated within the falling system, this slowing of time is not perceptible. Quite the contrary, in the right conditions, even the passage through the event horizon can be imperceptible to that observer.

In better understanding what's going on in the vicinity of a black hole, surprisingly, it's Hollywood that helps us out. In fact, the scientific advisor for the famous film *Interstellar* was none other than theoretical physicist Kip Thorne, who received the Nobel Prize in 2017 for the discovery of gravitational waves. Kip has devoted much of his life to the study of general relativity; he is a highly original character, curious explorer of the universe but also a scholar interested in developing the media's enormous potential for disseminating scientific knowledge. I met Kip in 2016, during the *Gravity* exhibition created by the Italian Space Agency (ASI) at Rome's MAXXI museum. Rarely have I had the pleasure of talking with someone who was so approachable and easygoing but at the same time so profound. We talked at length about his role in the preparation of the film and he confided that he had worked on it for almost twelve years, pursuing a fixed idea: making the effects of general relativity visible, using the extraordinary communicative power of cinema to make the general public aware of the results of Albert Einstein's monumental work. Anyone who has seen the film certainly cannot forget the images of the protagonist, astronaut Joseph Cooper—in his spacecraft *Endurance*—orbiting around the giant black hole, Gargantua, with its mass of 100 million suns and rotating at 99.8 percent the speed of light. It is just as impossible to forget the twenty-three years gained by Cooper relative to his daughter Murphy during the few hours he spent on the first planet upon which he landed, "Miller's" planet, which orbits—with the resulting gigantic tidal waves—in the vicinity of the black hole. The years gained near the black hole are the proof of time slowing down in the vicinity of the event horizon and, in the context of the film, create the opportunity for a moving final scene in which Cooper again meets his daughter, now older than he is, on her deathbed and surrounded by her children and grandchildren.

Thanks to Kip and his team of scientists, the first visual representations of the complicated "mirages" produced by a black hole relative to the background of the stars were developed. The path of light passing near a black hole can be bent to such an extent that a ray of light can make one or more orbits before reaching our eyes. This is how every single bright object in the background is multiplied and distorted countless times, with a jaw-dropping effect that was never before calculated by anyone. Anyone who wants to learn more about black holes, without having to become a theoretical physicist or wrestle with Einstein's equations, would get a lot out of reading Thorne's book *The Science of Interstellar*. In this book Kip also takes us inside Gargantua. As I mentioned, here we have only the equations of general relativity to understand what is going on; experimental verification is not possible and it is best not to rely on intuition, considering that space-time behaves in a completely unpredictable way. Kip himself tells us that some of what we see in the film, particularly the scene of the *tesseract*—the four-dimensional, cubic space from which Cooper is able to communicate a coded signal to his now adult daughter—is all the invention of special effects, the only way to portray something that, in truth, we don't understand at all.

With that caveat, I repeat that the film, and more importantly the book, can help us understand some of the most significant aspects of what we are discussing. First of all, as we have said, for the external observer looking at a phenomenon taking place in the vicinity of a black hole, time stops on the event horizon. The same is not true for those who are approaching or even crossing over it. Being in free fall, the presence of intense gravity is not perceptible, whereas there are strong tidal effects, which is to say the difference in gravitational force between two different points on the same object, for example, the human body. If the black hole is large enough—in *Interstellar* it had a diameter equal to the Earth's orbit—the tidal effect is negligible, similar to what we feel on the surface of our planet. However, if the black hole is small, the effect can be so intense and violent that it causes the immediate *spaghettification* of any unfortunate explorer: the atoms of the feet and those of the head go off on different trajectories and nothing can be done to prevent it. So, if you have to fall into a black hole, I recommend you choose the largest one possible!

Once the event horizon is crossed, free fall continues to the center of the black hole, but it isn't a point like any other: it is a *singularity*, an area where the effect of gravity grows indefinitely, limited only by the effects—not well understood, as I have already mentioned more than once—of quantum mechanics. How long does this fall last? Well, you have to consider the distance to travel, the speed, and the gravitational acceleration. It could take days or even months. In reality, it's not that simple. As the attentive reader will have begun to suspect, black holes are a continuous source of surprises. The fact that what falls into them disappears from sight for those who are on the outside does not mean that the information corresponding to what is inside disappears from sight for an observer who has ventured beyond the event horizon. On the inside, light moves in all directions, with oscillation frequencies and paths altered by gravity, but it does not disappear. Quite the opposite: as Thorne describes in his book, in addition to the central one, there are at least two other intermediate singularities—precisely related to the dynamics of light inside a black hole—one of which was discovered while *Interstellar* was being written.

So, let's leave the black hole explorer to his fate. In the end, Cooper is able to escape from Gargantua only with the help of extradimensional entities that are neither specifically described nor predicted by Einstein's relativity, but this is, of course, artistic license. What does exist is an extremely important issue that has to do with the disappearance of information within a black hole relative to an external observer. Even a force as majestic as gravity has to come to terms not only with quantum mechanics—with the second law of thermodynamics, which predicts that the disorder in the universe, known as entropy, can only increase over time—but also with the conservation of fundamental quantities: energy, charge, and other quantities, one of the few but solid principles of physics that is impossible to escape. From this perspective, a black hole seems an exception to the rule: it is a kind of cosmic entropy vacuum cleaner. Anything, however it enters the inside, appears only as mass, angular momentum, and electric charge from the outside. It would seem to be the ideal solution for toxic and radioactive waste, were it not for the fact that it is a little bit cumbersome to manage on Earth. This problem—clearly at odds with the fundamentals of physics—caught the attention of the scientist Stephen Hawking. In 1973, Hawking learned from two Soviet colleagues, Yakov Zeldovich and Alexei Starobinsky, that a rotating black hole can emit energetic particles. Shortly thereafter he was convinced by his own calculations that a nonrotating black hole is characterized by this effect.

Such radiation would solve the problem of entropy because, by causing black holes to lose energy, it makes them "evaporate," returning to the universe the information that was temporarily subtracted from it. The interesting thing is that, for solar-sized black holes, this time span far exceeds the current life of the universe and, as the mass increases, that time extends further. So, black holes are efficient but temporary deposits for entropy, which will return it to the cosmos, but over extremely long time spans. One thing is certain: physics is very subtle and small details often turn out to be fundamental. The radiation predicted by Hawking is one of his most important discoveries; the formula for the temperature of black holes is carved on his tombstone. This radiation has yet to be detected but is actively being sought with satellites observing the universe, in search of anomalous energy emissions from areas of the universe where there is nothing—pardon!—where there is a black hole.

21

TOWARD A NEW ASTRONOMY

Spring 2012, in the Tuscan countryside, near Cascina, in the province of Pisa. I was going to one of the meetings that had been organized to evaluate the progress of a European group of scientists engaged in the development of a highly sensitive interferometer, an instrument designed for researching gravitational waves and based on the interference of two laser beams traveling along perpendicular paths. I got off state highway 206 and took a secondary road, one that runs through the open countryside, alongside the Arno river's floodway canal. After a long, straight stretch, you take a curve and suddenly one of the world's most sophisticated astronomical instruments appears: Virgo, in the eponymous research laboratory of the INFN (National Institute for Nuclear Physics), managed by Italy in partnership with France and the Netherlands. This laser interferometer, formed by two three-kilometer-long arms, is part of the LIGO-Virgo collaboration, which would become famous on February 11, 2016, when the discovery of gravitational waves was announced.

The story of gravitational waves is a story of farsighted people, starting with Albert Einstein, whose general theory of relativity, published in 1916, predicted them a century before their detection. However, it wasn't an easy prediction. As early as 1905, Henri Poincaré had already proposed their existence, analogous to what happened with the emission of electromagnetic waves by an accelerated charge. However, when he was formulating his theory, Einstein was perplexed by Poincaré's idea, given that it implied that in gravity, mass has only one sign and there was no concept of a gravitational dipole, a configuration of electric charges characteristic of electromagnetism, formed by positive and negative charges. Nevertheless, Einstein developed a calculation, based on a particular approximation, from which the existence of gravitational waves was derived. That prediction was criticized by Eddington in 1922, while in 1936 Einstein himself, together with Nathan Rosen, submitted an article to the scientific journal *Physical Review*, in which the nonexistence of gravitational waves in the general theory of relativity was demonstrated. It was thanks to the intervention of an anonymous reviewer and Einstein's assistant, Leopold Infeld, that an error was identified, the article was withdrawn, and then resubmitted with the opposite conclusion! Quite some time later, in the early 1970s, the American Joseph Weber announced the detection of gravitational waves using resonant aluminum bars, the results of which later turned out to be spurious. We would have to wait until 1974 for the indirect detection of gravitational waves: Russell A. Hulse and Joseph H. Taylor observed how the orbital period of a binary-pulsar system changed over time and concluded that the data corresponded exactly to the predictions of general relativity. In 1993, they won the Nobel Prize for this discovery. But the holy grail, the direct detection of gravitational waves, was still to come; we had to wait for the development of a new type of instrument: large-scale laser interferometers.

The construction of Virgo owes a great deal to an Italian researcher, now deceased, Adalberto Giazotto, who in 1980, at the age of forty-one, returning from a period spent in England, decided to embark on a project considered fascinating but extremely difficult: the direct detection of gravitational waves. In Rome, Giazotto had worked with Edoardo Amaldi, one of the fathers of Italian physics, founder of the INFN, and a pioneer in the search for these signals.

Like Weber, Amaldi was also developing a technology based on resonant aluminum bars at cryogenic temperatures, which were suspended from very thin wires to isolate them from any ground vibrations. The bars resonated at frequencies of around 1,000 Hertz and were therefore designed to couple with gravitational waves oscillating at those frequencies. However, these frequencies are quite high, due to a gravitational process involving large masses, and consequently the sensitivity of the bars was limited. These detectors would have detected only extremely violent, and therefore very rare, events. On the other hand, cryogenic bars were, at the time, the best technology available to the scientific community. Giazotto began looking at a different measurement method, pursuing decidedly lower frequencies, in the tens of Hertz, an area in which the expected signal is both more frequent and more intense, but also one in which the background noise from Earth-based vibrations is higher. In order to move beyond this impasse, Giazotto developed superattenuators, a highly sophisticated set of pendulums and springs that were able to insulate the elements of the interferometer from any earthly seismicity.

I met Adalberto at the end of the 1970s. We were at the same laboratory, the INFN section of S. Piero a Grado, near Pisa. I was there working on my degree and then my specialization in the field of elementary particles with experiments at CERN in Geneva. He was there developing, step by step, the project that would become Virgo twenty years later. He was an enthusiastic, determined person, highly respected on a scientific level but without much of a following in the research community. The fact is that, without him, the INFN would never have committed to Virgo, considering that all of the resources for this type of research had already been invested in the cryogenic bars developed by Amaldi and his group of researchers in Rome. They had not yet realized that the limited sensitivity of these instruments meant that they would never have detected a gravitational signal; in fact, they were decommissioned only a few years ago.

I had the opportunity to collaborate with Adalberto several years later, between 2010 and 2014, when, as president of the INFN National Commission that financed the experiments in the astroparticle sector, I found myself closely following Virgo's enhancement, Advanced Virgo, which followed almost eight years of study and development carried out with the first version.

The advanced version anticipated the sensitivity would increase by a factor of 10. With a little luck, it could be enough to detect the first events: we were talking about an investment of around 22 million euros, in addition to the initial investment of 77 million, spent on the construction of Virgo. These may seem like significant sums, but they pale in comparison to the investments made by INFN during the same period for the accelerators and detectors that, in 2012, led to the discovery of the Higgs boson at CERN in Geneva. Meanwhile, a similar project, called LIGO, was being developed in the United States. It was composed of two interferometers, one of which is double, located at a distance of 3,000 kilometers. It must be said that the American investment in LIGO—even taking into account the expenses for personnel, which in the United States is included in the overall estimates of project costs—was more than five times the European one, demonstrating that the scientific community across the Atlantic was taking the challenge of the century in the gravitational sector very seriously. In any case, toward the end of the first phase, the decision was made to initiate a large-scale collaboration between Europe and the United States. It was convenient for everyone. In fact, it was clear from the calculations that instrumental sensitivity increased with the number of different interferometers and the distance between them. When the two advanced interferometers were switched on, LIGO in September 2015 and Virgo in February 2017, they quickly delivered results. In late 2015, the first gravitational event was detected by the two LIGO interferometers, which also detected other events in the following months, until, in August 2017 there was the first event that was also detected by Virgo, thus involving all three interferometers that were part of the collaboration.

But let's go back for a moment to 2012. At that time, the race to create the advanced version was in full swing; everyone understood what was at stake, but at the same time the enormous disparity in investment between Virgo and LIGO was evident. Unfortunately, it was too late to change things. We were faced with a typical problem of scientific sociology: the community involved in Virgo (a few hundred people) was small compared with the one involved in the LHC experiments at CERN in Geneva (thousands of researchers). Consequently, economic support at the European level was limited, also due to the high risk of not achieving any result, because of the extreme difficulty of measuring such faint signals. I realized in that situation, barring miracles, we would have lost the race for the Nobel Prize, despite Italy having reached, thanks to Giazotto and his team, an extraordinary degree of competence. Back then, following the closure of the new but never completed large Superconducting

Super Collider or SSC accelerator in Texas, the American science community could focus on the quest for gravitational wave detection. This was another reason they had been able to bring more substantial resources to bear on gravitational waves and were aiming to lead the competition. Fortunately, the agreement between the two collaborations was maintained and the Europeans didn't miss the train, which then led to magnificent results, even if the Advanced Virgo detector was not yet operational in the decisive year in which the first gravitational wave event was detected. Today Europe is proposing the next large interferometer, called the Einstein Telescope, a futuristic instrument placed underground and capable of further increasing the sensitivity of the detectors. There is still some bitterness about not having been able to grasp the magic moment at the end of the 1990s, of not being able, after thirty years of extraordinary preparatory work, to promptly and adequately recognize and support the extraordinary vision of men like Adalberto Giazotto.

But what is Virgo's laboratory like? At first glance, it resembles other research centers like Geneva's CERN, which you may have seen in newspapers or on television: low-rise buildings filled with offices, workshops, labs, computing centers, and meeting rooms. But, little by little, you might notice that there is something different about Virgo. Activities at a high-energy physics laboratory are centered around the functioning of one or more accelerators. These are instruments—as the name implies—designed to accelerate particles to the highest possible energies. They are formed by large magnets so that the particle beams are forced to travel in a circular orbit, and by radiofrequency cavities operating at high voltage to accelerate the beams' particles. An accelerator consumes prodigious amounts of energy, so the electricity bills for this kind of facility are a major expense. For instance, this is why CERN doesn't operate during the winter. At some points along the orbit, the beams are made to collide in order to create secondary particles whose collisions are "photographed" by complex detection instruments, sometimes as large as a five-story building, as in the case of the ATLAS (A Toroidal LHC Apparatus) and CMS (Compact Muon Solenoid) experiments. In the control room, the scientists monitor the frequency at which the collisions occur; the more collisions per second there are, the brighter the accelerator and the faster the data acquisition.

In the case of gravitational waves the focus shifts: the aim is to make two laser beams, traveling in perpendicular directions, take the longest possible path between two mirrors (minimizing any disturbance factor), before bringing them together. How is it done? Passing along two 3-kilometer-long, perpendicular vacuum tubes, light bounces off ultraprecise mirrors dozens of times before being sent to the "detector," a photodiode where the two beams are recombined. If the beams arrive in phase, the light intensity is at its maximum; if they are in counter phase, a dark area is created. In the control room the scientists observe any changes in the patch of light projected on a large screen.

The atmosphere at Virgo is a bit surreal. At an accelerator like the one at CERN, dozens of control and power systems work in a perfectly synchronous way to ensure that—as previously mentioned—very violent collisions between energy particles occur. At Virgo, they try to make sure that nothing happens to the beams of light. This is the only way that the faint signal from a distant cosmic cataclysm, carried on gravitational waves, can be identified and distinguished from the effects of background noise. When it comes to sounds, we could say that, while at CERN they are working to create big, loud explosions, at Virgo they are working to create perfect silence and listen to the whispers of the cosmos.

Then there was September 14, 2015: the dawn of a new astronomy. What happened that was so momentous?

As I hinted at earlier, LIGO's analysis system identified, in both of its interferometers, a clear oscillation of the patch of light in the detection zone. It was an unmistakable signal, which the modeling traced back to the merger of two black holes in a distant galaxy—one of about 35 solar masses and one of about 30 solar masses—into a single black hole of around 63 solar masses. The effect of this cosmic cataclysm, which in the final milliseconds released an amount of energy that exceeded the brightness of the entire universe for an instant, caused a variation in the length of LIGO's arms that was about one-thousandth the diameter of a proton, which, when brought back to the size of our galaxy, is comparable to the width of one hair. Since then, a few dozens of events have been detected, including one related to the merger of two neutron stars. Compared with the merger of black holes, merging neutron stars produce exceptional fireworks of visible light, observable using traditional astronomical instruments, both ground-based and space-based. One of the most interesting aspects of observing gravitational waves is the exceptional detail that can be extracted from the signals that reach the laboratory. In fact, the general theory of relativity gives precise indications about how these phenomena occur, and the modeling corresponds quite well to the observations. The mass and type of the merging bodies, their state of rotation, their speed and relative position are all parameters that can be determined. With the current level of sensitivity, when the interferometers are working, there is a detection rate of about one event per week. To date, there are dozens more signals awaiting confirmation. Gravitational wave detection has opened a new chapter in astronomy that can, in terms of importance and without a shadow of a doubt, be likened to Galileo's first observations with what was then a revolutionary instrument, the telescope. The waves allow us to observe an otherwise dark part of the universe, dominated by black holes and neutron stars, and to study in detail phenomena that may last just a few seconds or minutes, depending on the characteristics of the initial system. The observation of the light emission associated with the neutron-star events marked the beginning of multi-message astronomy, in which the same phenomenon can be observed through two types of waves: gravitational and light.

Among other things, these observations make it possible to take precise measurements of the relative speed of the two types of waves, which is identical, except for a very small degree of experimental uncertainty. We can also understand how the extreme conditions created by the merger of neutron stars are capable of producing heavy nuclei, such as gold or other elements heavier than iron and therefore not possible to synthesize within stars. It is estimated that in one of the observed events, a collision between two neutron stars, enormous quantities of heavy metals—gold, platinum, uranium—were produced. Of gold alone, it would have produced a volume comparable to that of the Earth. With the observation of gravitational waves, black holes have become the subject of intense experimental study. For example, in April 2017, a group of scientists from the Max Planck Institute used the observational capabilities of eight large radio telescopes to create a super telescope with an observational base as wide as the Earth, and with a corresponding angular resolution of about

20 microarcseconds. Thanks to sophisticated analysis, performed with the use of a supercomputer, which took about two years, it was possible to create a direct image of a huge black hole, located in a galaxy 55 million light-years from Earth. Another extraordinary advance in twenty-first-century astronomy.

22

TOWARD THE INFINITELY LARGE

Let's resume our journey toward the infinitely large. What can we say about the current and future dimensions of the universe? Considering that the universe has been in existence for about 13.8 billion years, and that it is constantly expanding, the part of its diameter "accessible" to observation is on the order of 93 billion light-years, about 10^{27} meters. Half of this diameter corresponds to the maximum distance that a photon emitted at the beginning of time can have traveled before reaching us. As we saw earlier, the observable radius of the universe is quite larger than that calculated by multiplying the age of the universe by the speed of light. The reason for this is that the space-time metric has continued to expand and the speed of light is finite. This means that today we can observe "ancient" light coming from galaxies that, in the meantime, are moving away from us so fast that they are now invisible.

Taking into account the continuous expansion of the metric between each pair of points placed at great distances, whose relative speed is determined, as we have seen, by Hubble's constant, it is easy to see how the size of the observable universe changes over time. Each passing moment we lose sight of distant galaxies that used to be visible. At the same time, the edge of the visible universe increases at a little less than the speed of light, incorporating light radiation that, until an instant before, was not part of *our* observable universe. At the present time, starting from the typical dimensional scale of a human being—using the meter as our benchmark—there are about twenty-seven orders of magnitude that separate us from the maximum distance accessible to us in some way. This is a huge number, but in all probability infinitesimal compared with the true size of the universe.

By running through the various time and space scales of our universe, we have understood some of its properties, which, as far as we know, are the same everywhere. The traces of the primal fluctuations, over time, led to concentrations of dark matter, which in turn led the immense initial cloud to fragment into countless structures. The latter then led to the birth of black holes, stars, and galaxies. The continuous work of gravity, accompanied by the other fundamental interactions, has enriched the cosmos with heavy elements, which were not present in the immediate aftermath of the Big Bang. Almost 14 billion years after its origin, the universe is still far from having reached its equilibrium. Spatial expansion, in particular, is continuing. At the age of around 10 billion years, the density of dark energy surpassed that of matter, and today, at cosmological distances, the expansion of the metric is accelerating, probably due to the effect of dark energy. This accelerated expansion seems to continue for a time that exceeds that of the current age of the universe.

What can we say about the part of the universe that is not accessible to us and therefore not observable? Not very much. We have to limit ourselves to some speculation, since, beyond the range of observability, we are not causally connected with the rest of the cosmos. The measurements of cosmic microwave background radiation, which allow us to observe the properties of the universe at the time of recombination, that is, 379,000 years after the Big Bang, when the temperature dropped and allowed the formation of helium and hydrogen atoms, doesn't give us much information about the true size of the universe, even if they make it possible for us to determine the average curvature of the accessible part. The latest measurements from the ESA Planck satellite—launched in 2009—return a zero mean curvature value, with an error of five parts per thousand, so they are not very helpful from this perspective. Based on what we know today, it is reasonable to think that the part of the universe we can observe is only a small part of an exceptionally vaster universe. But, above all, it's interesting to ask ourselves a question.

Is ours really the only universe that exists? Or are there one, two, or a thousand others? We don't have the answer to that. Or better, in this case, more than in any other, we are destined to remain in the realm of speculation. If the origin mechanism we discussed in the first chapters is correct, we are faced with the very strange situation in which the universe emerges out of nothing without having expended any energy for its birth. According to the theoretical model that is currently the primary benchmark, in the earliest stages of the Big Bang, dominated by quantum fluctuations, the characteristics of the fundamental fields were defined through the breaking of symmetries present in the interactions between elementary particles but valid only at very high energy. It's as if, at the beginning of time, we had a circular table, perfectly set, ordered, and symmetrical, but, after the guests' arrival, each in different clothes, with individual habits and physical characteristics, most of these symmetries were lost. Depending on the parameters defined in this symmetry-breaking process, a universe with specific characteristics can develop, more or less rich and structured than ours, in most cases unable to form particles, atoms, and molecules, lacking, in addition, the complexity necessary for the development of life. This is how we get to the formulation of the Anthropic Principle, according to which the universe in which we live is what it is because only this particular expression of the laws of physics has the potential to develop sentient life capable of observing and reasoning about such a universe. According to this line of reasoning, there are an infinity of other universes, completely disconnected from ours, which develop according to different laws, established in the Big Bang stage. Universes in which the laws of quantum mechanics, entropy, and relativity are (probably) the same, but in which the mass and charge of the elementary particles, the value of the Planck constant, the intensity of the various forces, and other conditions are not the same.

It is a hypothesis that has a certain fascination. On the other hand, we have little more than methods of analysis and deduction at our disposition when trying to understand if there are other universes. This is not, of course, a completely new idea; quite the opposite. Let's think, for example, about Giordano Bruno's bold reflections—which I have cited in this book for a reason—on the infinite worlds in an infinite universe. The fact remains that, in order to explain our existence, microscopic dust in an immense universe, we have not been able to do better than to raise the ante by introducing an in(de)finite number of universes, only one of which is able to support us, for a period equal to the cosmic blink of an eye. Is this convincing? Yes and no. One naturally wonders if there are not more "economical" conceptual scenarios, capable of describing the universe in which we live without disturbing an infinite number of them, completely unreachable for us. It's a bit like when, after having learned to use numbers to do the four operations and to resolve ordinary, day-to-day, limited problems, we find ourselves exposed to the vertigo of infinity in mathematics. In physics, the question is even more subtle: it would seem that we are dealing with infinite zeros, only one of which has anything to do with us. All of the others can only be contemplated in our minds.

23

TOWARD THE INFINITELY SMALL

In our journey through the cosmos, we have arrived at the far reaches of the universe, borders that exist because of a limit on observability, currently considered unsurpassable: the speed of light. Now we will begin pursuing a path, no less fascinating, toward the infinitely small. We will discover that, also in this case, at a certain point we must stop, when confronted with an obstacle of an experimental nature, due to the limitations of our most advanced observational instruments: particle accelerators.

In 1610, when Galileo observed that the surface of the Moon "is by no means endowed with a smooth and polished surface," as claimed by "all astronomers and philosophers," but, on the contrary, is "rough and uneven and, just as the face of the Earth itself, crowded everywhere with vast prominences, deep chasms, and convolutions," therefore covered with mountains and valleys, it opened the doors to a consideration that, for the first time, makes a connection between things on a human scale and the immensity of the cosmos, which had always been an unreachable, mythological place. In the following decades, with the invention of the microscope, the dimensional scale would be even further extended toward the invisible world of the very small. However, from a strictly logical point of view, realizing that the matter of which the planets are made is the same as that which surrounds us and as that from which we are also made represented a revolution of inestimable significance.

Now we are used to the idea that the laws of physics are valid throughout the universe, formed by the same type of matter, but it was only in
the early 1900s that we started to understand how the atomic and subatomic microcosm is made. It is an invisible world, governed by quantum
mechanics, which can be studied only through ingenious instruments
capable of providing measurements that, once interpreted, allow us to
understand the properties of the objects observed. It should be emphasized that, from an experimental and conceptual point of view, dealing
with stars and planets is very different from dealing with atoms, nuclei,
and elementary particles. In the microcosm we move forward like blind-
folded players batting at a piñata with a stick, trying to break it to reveal
the hidden treasure. Our "sticks" are particle accelerators: the higher the
energy of an elementary particle, the more it becomes an effective probe
for studying the microscopic world. In order to do this we can use natural
accelerators, like cosmic rays accelerated by magnetic fields in the depths
of space, or artificial ones, like the highest-energy accelerators at CERN
in Geneva. What we learn from these experiments concerns the fundamental structure of the physical world; its elementary, indivisible components; and the forces that regulate their behavior.

It is thanks to cosmic rays that a series of new elementary particles
were discovered, at the beginning of the twentieth century, leading to a
new field of research: subnuclear physics.

In the second half of the twentieth century, increasingly large and
complex artificial accelerators took over, culminating with the gargan-
tuan accelerators, which have enabled the discovery of the sixth quark,
the electroweak bosons, and the Higgs boson. For all intents and purposes, a particle accelerator is a powerful microscope based on quantum
physics. In the extremely violent collisions between highly energetic particles, a virtual state—similar to the quantum fluctuations that characterized the Big Bang, but with a positive total energy—is created for a brief
instant. This state decays immediately, giving birth to new particles that
can be different from those that initially collided. By supplying a lot of
energy to the initial particles, it is possible to create other particles with
very large mass and a very short half-life. For example, the electroweak
bosons and the Higgs boson, discovered at CERN, have a mass almost 100

times greater than that of a proton, while the sixth, or *top* quark, discovered at Chicago's Fermilab, has a mass almost 200 times greater.

We are talking about an effect that is entirely due to quantum mechanics and to the equivalence between mass and energy found in special relativity. Just to be clear, it is as if we threw cherries at an apricot tree and grapes popped out. In the microscopic world, it's possible to provoke transformations between different types of particles.

After nearly a century of research, the Standard Model of elementary particles and fundamental interactions, which I mentioned in the previous pages, emerged. So, what are we talking about? For subnuclear physics it's the analogue of Mendeleev's periodic table: a sophisticated, elegant schema that describes the building blocks of matter and the way in which they interact with each other through the three fundamental forces: electromagnetic, weak, and strong. The Standard Model does not deal with any particles of dark matter or dark energy present in the vacuum; it only describes and systematizes less than 5 percent of the matter in the universe. In order to describe any other types of particles, such as dark matter particles, the Standard Model has to be expanded. This does not mean that it isn't an extraordinary result. The ideas and theoretical tools that accompanied its development will almost certainly prove useful for identifying directions for future research.

What does the Standard Model say? First, it separates elementary particles into two types: fermions and bosons. The names originate with two physicists: the Italian Enrico Fermi and Satyendra Nath Bose of India. Fermions are the building blocks of matter; bosons are the intermediaries between fundamental forces. Both types of particles have a fundamental characteristic that makes them distinguishable: they carry a certain elementary quantity of angular momentum, known as *spin*. For all intents and purposes they spin around like tops, even if, being quantum tops, the spin is quantized in semi-integer units. Fermions have a half-integer spin $(1/2, 3/2, \ldots)$ and bosons a whole one $(0, 1, 2, \ldots)$.

Elementary fermions are grouped into three *families*, each one made up of four different particles: two *leptons*, one charged and one neutral, and two *quarks*, one with an electric charge of 2/3 and one with a charge of −1/3. For example, the first family contains the *electron–neutrino* electron pair and the *quark u–quark d* pair. As previously described, quarks cannot be observed on their own; they form more complex particles such as *mesons*, made up of quark–antiquark pairs, or *baryons*, composed of three quarks.

The particles of the Standard Model that are stable or have very long half-lives play a precise role in the construction of the universe. These include neutrons and protons, the constituents of the atomic nucleus, which are examples of baryons. Or the electrons that orbit the nuclei, neutralizing its electric charge, forming atoms. Neutrinos, for example, play an important role in the nuclear fusion reactions that feed the energy of the stars. When it comes to force mediators, we have the *photon*, for electromagnetic interactions involving charged particles; the three *intermediate vector bosons*, $W^\pm$ and Z^0, for weak interactions (involving all of the elementary particles); and the eight *gluons*, for strong interactions (involving only quarks). All of these bosons, which mediate the elementary forces, have a spin of 1. In contrast, the mediator of gravitational force, the *graviton*, has a spin of 2. It should be noted that gravity, the fundamental interaction that involves all of the elementary particles, is not part of the Standard Model due to the difficulties encountered in harmonizing this force with the other three in a single theoretical framework. Finally, there is a special boson with a spin of 0, the Higgs boson, responsible for the mechanism that allows all of the elementary particles to acquire mass. To our current knowledge, all of the elementary particles of the Standard Model are points of matter without internal structure, characterized by some fundamental properties such as charge, mass, and spin. From an experimental point of view, we can say that, if an electron has an internal structure, it is smaller than 10^{-18} meters.

It took decades to achieve the formulation of the Standard Model, with its simplicity and elegance. It will be useful to retrace the decisive steps in its construction, because it is a textbook example not only of the way in which nature hides its fundamental properties from our eyes but also of how much intelligence and tenacity is required of researchers who want to remove this veil of mystery and move toward a deeper level of knowledge. The first challenge was understanding atomic structure. It took Ernest Rutherford's experiments from 1908 to 1913 to understand that the atom had a compact, heavy core, positively charged and surrounded by light, negatively charged electrons. Without quantum mechanics, developed around the same time, we would not have been able to understand the properties of atoms and their subatomic components. Contemporaneously, experiments with cosmic rays and those done with the first accelerators led to the discovery of a number of particles that were *unstable* and therefore unobservable in ordinary matter. Further research led to the astonishing discovery of *antimatter, antiparticles* with characteristics opposite those of the known particles, like the positron (the antielectron) and the antiproton.

Quantum mechanics—initially nonrelativistic, then developed to include special relativity and therefore the equivalence between mass and energy—provides a very powerful method for the theoretical analysis of the properties of elementary particles and the force fields that cause them to interact. The extensive period during which this theory was being developed was an exciting one: it laid the foundations for a great deal of research, which then came together with the study of the first moments of the universe. As more powerful accelerators were built, new particles and new phenomena, always very unexpected, were discovered. I'm thinking about the violation of some symmetries considered fundamental. One example is parity, or P-symmetry: in the macroscopic world two objects with symmetrically mirrored properties can always be realized, while in the microscopic world this is not always true. Neutrinos, for example, are only left-handed, think of it as if they were rotating in only one direction with respect to their velocity direction; they do not have a right-handed mirror particle, while electrons do have both left and right components. Another example is time-inversion symmetry, or T-symmetry: in the macroscopic world a film projected backward, for example, of a broken vase that is reassembled, might make people laugh but it also corresponds to something that could theoretically happen. However, in the microscopic world, some reactions between elementary particles are influenced by the direction of time's arrow: they happen at a different rate going forward or backward in time. Several of these discoveries have garnered Nobel Prizes.

It was not quantum mechanics that permitted us to predict how many types of particles existed and what their properties were. It was the experiments, carried out by increasingly powerful accelerators, that systematically explored increasing scales of energy, discovering the masses of the different elementary particles. Most discoveries were a genuine surprise, but each time we thought it was the last word on the subject. For example, when it was understood that there were three types of quarks and theories were developed that could explain all of the nuclear states formed by three types of elementary particles. Except the model was then abandoned when, unexpectedly, a new elementary particle was discovered, the J/Psi, which was immediately interpreted as formed by a fourth type of quark and the formulation of families constituted of pairs of particles began to be affirmed. The discovery of the fourth quark, dubbed c, as in *charm*, caused a big stir and was published simultaneously by two competing groups in November 1974: that of Samuel Ting in Brookhaven, New York, and that of Burton Richter in Stanford, California.

A number of other important discoveries happened in those years. In 1977 Leon Lederman of Fermilab discovered the still heavier *upsilon* meson, interpretable as a bound state of the fifth quark and its respective antiquark, which was called the b, or *bottom*, quark. At this point, everyone was expecting the sixth quark, t, or *top*, to be discovered at any moment, but no one expected that it would have substantial mass—around 180 times the mass of the proton—and that it would be twenty years, which is to say 1995, before it was discovered at Fermilab.

The history of the J/Psi and the November revolution had a decisive influence on my scientific development. I began my thesis in 1978, at CERN, working on the experiment of recent Nobel laureate Sam Ting. Ting was a young Chinese-American professor from MIT, known for his determination and ability to go against the trend; a person with an outside-the-box character, he inspired the figure of the Machiavellian Chinese physicist in Daniele Del Giudice's book *Lines of Light*. In 1978, I would never have imagined that our collaboration at CERN would last for almost forty years and would continue into space, searching for antimatter in cosmic rays. We'll talk more about that later.

24

BIG SCIENCE

CERN—it must be remembered—played a decisive role in elementary particle physics, and today, it is the world's most important laboratory operating in the field. Founded by a group of European scientists, including Italy's Edoardo Amaldi and France's Pierre Auger, it symbolized the rebirth of Europe's community of nuclear physicists following the disasters of the Second World War, the diaspora connected with the Manhattan Project, and the resultant intense militarization of this sensitive scientific field. For tens of thousands of scientists from all over the world, CERN is synonymous with research into the fundamental laws of nature, research carried out at the highest levels and with the most sophisticated experimental and theoretical tools. It is no coincidence that the laboratory has become so well known that it inspired part of Dan Brown's bestseller *Angels and Demons*.

Because of needs related to the complexity of the instruments used, work at CERN is organized into international teams. It makes me smile when I think that, in the late 1970s, a big team would have been composed of thirty people and an experiment would have lasted three to four years. Today, LHC experiments involve around 5,000 people and can continue for decades.

Just forty years ago, the first sight of the immense laboratory, crisscrossed by streets named after famous scientists and operating around the clock, left a deep impression on me. It is a place where researchers met and talked with each other over a cup of coffee; they filled the vast meeting rooms to listen to seminars or to organize their group projects. It's still that way today, but the development of the internet and social media has changed the way people participate in common activity.

For me, CERN's library, located in the center of the lab and open 24/7, is unforgettable. I have always loved books, the universal archive of humanity's knowledge. It must be said that the books fundamental to the scientific field were (and still are) almost always in English, and therefore not so easy to come by in Italy. They were all at CERN, always available, at any time. Remember that, at that time, the World Wide Web did not yet exist—it was, in fact, invented at CERN in 1989, by Tim Berners-Lee—the internet was in its infancy, and there was no Amazon or any other web-based retail giant. I remember spending my nights in the library, sometimes sitting on the floor and surrounded by books, one open on top of another, in order to compare texts and references in real time. It was like having an analog computer with a lot of open windows and being able to jump from one PDF to another. Today this is normal; with a simple click all kinds of information is available whenever you need it. Back then, quite the opposite was true. We had to work hard to find the information we needed most. That has remained my ideal of study and research. For years, every time I went to the United States, I bought suitcases full of books to bring back to Italy. Important books were bought as a point of reference. Many would never be read and consulted only briefly, but it is a good thing to have the classics on your bookshelves, physical objects that link you to the authors, who in a distant time and place produced a cultural contribution that has taken on a universal value. These are the building blocks of our civilization. Without books, the memory of what we are disappears and, without memory, we become like mindless moths.

I often reflect on how things have changed since the invention of the web and the capillary spread of the internet. The diffusion of knowledge has become exponentially faster but, at the same time, more specialized and fragmented. Today, wherever I happen to be, I can do what in those years I could do only at CERN; all I need now is a computer and a good internet connection. In truth, I have to admit, whenever I need to understand something thoroughly or when I am trying to make a new contribution to a scientific problem, I have to immerse myself in the flow of information and swim against the current until I get to the true source of knowledge that, even today, I find only in the calm reading of the right book or article, be it on paper or on the computer screen. Paradoxically, now that it is so much easier to access books and research articles, it is more difficult to explore them in depth by dedicating the necessary time to them. Virgil says that *otium*, or leisure devoted to study and contemplation, is a gift from the gods. In our times, the *otium* described by the Greeks and Romans is the most precious and rarest resource, one of the most sophisticated and creative acts that characterizes us as human beings. It is also necessary to do good research, but all too often this is misunderstood by those who plan or finance that research. This is especially true in recent years where bibliometric assessment and specialization hold unchallenged sway relative to the quality of the content. Complicating matters even more, I've witnessed, in my country in particular, a deep and increasing lack of responsibility in judging the scientific value of the research results, often entrusted to mathematical indexes rather than experts.

In the late 1970s, physics was not only abuzz over the quark model. The foundations were being laid for a very important discovery for CERN—it was a few years in coming but, thanks to the work of a volcanic Italian, Carlo Rubbia, and a refined Dutch particle accelerator physicist, Simon van der Meer, it did finally happen in 1983. The subject was the unification of the electromagnetic force and the electroweak force, an important part of the super force we saw at work during the Big Bang. There was a theoretical prediction that forecast three large-mass particles, intermediate bosons W^+, W^-, and Z^0, the analog of the photon for electromagnetic interactions. Detecting their existence was likely the most coveted scientific goal of the time.

As always happens in true research, there were clues but at the same time a fair number of uncertainties. Various indirect measurements indicated, with a certain degree of accuracy, that the mass scale of these new particles would be found to be around 100 times the mass of a proton; however, there were no accelerators capable of reaching this exceptionally high energy. Rubbia had proposed a brilliant plan to resolve the problem, modifying the Super Proton Synchrotron (SPS), CERN's most powerful accelerator. The idea was to add to the SPS proton beams an antiproton beam traveling in the opposite direction; in this way it would be possible to reach the energy necessary to create electroweak bosons.

But no one had ever made an intense antiproton beam. Since these are antiparticles, they do not exist in matter. They can be produced in accelerators but it takes a long time, dozens of hours, because, typically, only one antiproton is produced for every 100,000 collisions. So, it was necessary to accumulate the antiprotons produced until there were enough of them to produce a beam that could then be accelerated in the SPS. However, no one knew how to trap and store so many antiprotons for that length of time.

Van der Meer and Rubbia succeeded by inventing a new method based on a sophisticated technique that involved "cooling" the antiprotons in a special circular accumulator. This was the breakthrough that made it possible to modify the SPS, turning it into the first accelerator to use a beam of protons that collided with a beam of antiprotons.

Not only was Rubbia working on this new scheme for using the SPS; he was also developing the giant detector for the experiment that was supposed to detect the intermediate bosons. Called UA1 (Underground Area 1), it was based on an enormous, 800-ton electromagnet, inside of which was an innovative detector capable of observing all of the charged particles produced in the collisions between the two beams in detail. It was the first "multifunctional" experiment, able to study all of the details of the events produced. This was another way in which Rubbia demonstrated his vision and ability to innovate.

After graduation, I started my doctorate in Paris, and I found myself working on UA2, the competing experiment that would operate in another one of the SPS collision zones. It was a smaller experiment, without a magnetic field, directed by the French physicist Pierre Darriulat. It had been approved to test what UA1 discovered, even though we tried very hard to get there first and missed by just a hair. I had met Carlo Rubbia, not yet Director-General of CERN, when he was working as a professor at Pisa's Scuola Normale. He was a mythical figure for all of the physics students and his seminars were very well attended. His reasoning was quick and brilliant; his character was such that he gave no quarter and frightened everyone. At CERN I used to meet Carlo at the main bar and talk with him about the progress of stochastic cooling and the construction of the two experiments. This laboratory was (and still is) a place to engage with exceptional people, to learn, while on your coffee break, things that can change your own research.

The SPS Collider, fed with antiprotons, was brought online in 1981. After a first year of low-intensity data collecting, UA2 recorded an event that in its final state had two electromagnetic particles: two very-high-energy electrons with a combined mass that coincided exactly with the expected mass for Z^0. Unfortunately, the second particle was at the edge of the angular coverage of the apparatus, and thus we were not sure whether it had been measured correctly. At that point, UA1 still had not collected enough statistics. Nothing came of it and, over the next two years, UA1 collected enough statistics to be the first to publish evidence of the observation of weak intermediate bosons in early 1983. The Nobel Prize for Rubbia and Van der Meer followed in 1984 "for their decisive contributions to the large project, which led to the discovery of the field particles W and Z, communicators of weak interaction." In 1989 Rubbia became Director-General, at that time he launched the construction of the large accelerator that would later become the LHC.

CERN is a place that makes you think about what science is, how research works, and how discoveries come about. As Jared Diamond masterfully described in his book *Guns, Germs, and Steel*, society develops science, as well as other cultural endeavors, when it reaches a critical mass, making it possible for a small fraction of its members to carry out activities that do not offer an immediate return on investment and are not linked to daily survival but that, in the medium term, do provide advantages not immediately evident at first sight.

This is why such activities never take root in small tribal groups or agrarian societies; first we have to wait for the birth of cities, then industrialized societies, before we witness the development of science and research. I often wonder how many Einsteins, Newtons, or Edisons there have been in human history. People who were brilliant, visionary, and daring people but unable to develop their gifts because they were born at the wrong place or the wrong time. In this sense, driven by insatiable curiosity, CERN represents the maximum expression of the human ability to deal with seemingly useless aspects that later, in a completely unexpected way, change the world's destiny. Remember the extraordinary example of the World Wide Web, invented to resolve problems related to communication and data exchange between scientists working around the globe, and made available for free through the HyperText Markup Language (html) developed by Tim Berners-Lee? This is not an unimportant detail. Just consider the economic impact of the Web as a whole: even the most sophisticated analysts wouldn't know how to evaluate it. There are now billions of websites, and even the data on daily access at a global level has reached an astronomical number. Just imagine what would happen if there were, even for just a moment, a tax of a thousandth of a dollar (a pittance) on every click. It would very quickly generate several billion dollars, far exceeding CERN's annual budget. So, we could say that the development of laboratories like CERN, which are done entirely with public resources, has been amply justified by the benefits that all of humanity has gained, and is gaining, from a single invention like the Web—a development that, when the Geneva laboratory was founded, no one could have predicted, like so many other game-changing discoveries.

25

THE UNIVERSE AS A LABORATORY

In the previous chapters we saw how the macroscopic dimension of the universe is intimately linked to the fundamental properties of microscopic matter. So, the universe can be considered an immense laboratory in which matter and energy are continually subjected to the broadest range of experiments, often on time, energy, and spatial scales impossible to achieve on Earth. A textbook case is the study of gravity, comparatively easier to research by observing the cosmos rather than in the lab, due to the enormous masses involved. But even the study of elementary particles and other fundamental interactions can benefit from observation of the universe; this area of research is known as astroparticle physics.

We can say that astrophysics is to astroparticle physics as a chess match is to the individual pieces. Astrophysicists are interested in the processes that formed stars and galaxies, extreme phenomena such as black holes, neutron stars, and supernovae. Astroparticle physicists, on the other hand, use the information coming from the universe to carry out experiments similar to those done with the accelerators, albeit with very different instruments and methods. In fact, by carefully designing the accelerators, the desired experimental conditions are created and the laws of nature are systematically questioned in the specific context created in the laboratory. This is something that cannot be done if you decide to observe the universe. In that case, you have to arm yourself with patience and take what comes through cosmic radiation. In addition to visible light, the universe sends many other signals, in particular, light at frequencies that in many cases are not visible to the human eye or at frequencies that are not even able to penetrate our atmosphere. For this reason they have to be studied using satellites in space. Other "messengers" include the gravitational waves coming from remote areas of our galaxy and the extremely elusive neutrinos, which require detectors equipped with enormous volumes of ice, water, or other special materials. Then there are cosmic rays, stable charged particles produced outside our solar system, accelerated over the course of millions of years by variable magnetic fields, until they reach, in some cases, energies that no artificial accelerator will ever be able to match.

In order to fully understand how essential a role the cosmos has played as a laboratory for understanding nature's fundamental laws, let us look back for a moment to Newton's formulation of the laws of motion and the law of universal gravitation or the decisive insights derived from the results of astronomical research carried out by Copernicus, Brahe, Kepler, and Galileo. In fact, it is only in interplanetary space that bodies are sufficiently free of the effects of external forces to proceed in practically uniform motion for extended periods of time, as predicted by Newton's first law. In contrast, here on Earth, after a while everything stops due to the omnipresent effects of friction. Precisely by observing the universe, Newton was able to understand that what happens on Earth—the force of weight, the motion of bodies—is influenced by the same laws that govern the motion of the planets and the stars. This is how the law of universal gravitation was discovered. The famous anecdote of the apple, regardless of whether it actually happened or not, explains how Newton, at one point, realized that motion of the distant Moon is subject to the same law of motion as an apple in free fall toward the ground. It is precisely this intuitive leap that enabled Newton to derive and calculate the link between gravitational acceleration on Earth, the one that determines our weight, and the constant that appears in the law of universal gravitation.

Another example is the role of cosmic rays in the birth of elementary particle physics. Following the discovery of cosmic rays by Victor Hess in 1912, it became possible for the most skilled scientists to have access to new types of elementary particles created by the collisions of cosmic radiation with atoms in the atmosphere. The study of cosmic rays led to the discovery of numerous new elementary particles: the first antiparticle, the positron, was discovered in cosmic rays, as was the muon and the antimuon; charged pions; and the first particles containing the third type of quark, the strange, or *s*-quark, the K mesons and the Lambda baryon. Thus an entire branch of physics began thanks to the study of information provided by the cosmos. So, access to nonordinary matter—that which under ordinary conditions does not come into play in physics but which is very important to understanding fundamental interactions— was provided by particles that owe their energy to cosmic accelerators.

However, astroparticle research did not stop with the advent of artificial accelerators in the 1950s. An essential discovery in the 1990s was achieved by accurately measuring the flux of low-energy neutrinos coming directly from the Sun's internal nuclear reactions or from the Earth's atmosphere, constantly bombarded by cosmic rays. Data collected from two underground experiments, the first at Canada's Sudbury Neutrino Observatory and the second at Japan's Superkamiokande, led to the Nobel Prize awarded in 2015 to Arthur B. McDonald and Takaaki Kajita, for having clarified that neutrinos, which were thought to have zero mass, have a negligible mass—but not zero—and that the three types of neutrinos can oscillate between one state and another over the course of their existence (each of the three particle families in the Standard Model contains a different type of neutrino). Again, this is a discovery made by decoding a complex signal hidden in the radiation emanating from the cosmos and finding an effect that experiments at particle accelerators were unable to detect.

Therefore, the competitive collaboration between experiments with accelerators and those using cosmic radiation has been underway for more than a hundred years. And it is ongoing, with today's scientists using ever more complex instruments. While the largest accelerator ever constructed, the Large Hadron Collider at CERN, allowed the discovery of the Higgs boson, which we'll discuss more fully in the following chapter, instruments of enormous size—such as the Pierre Auger Observatory, a gigantic array of detectors spread across 3,000 square kilometers in the Argentine pampas, or IceCube, a cubic kilometer of ice in Antarctica equipped with strings of optical sensors—explore the celestial vault in search of new fundamental phenomena, using extremely high-energy particles from the depths of the cosmos. A century later, the study of cosmic radiation has retained its potential for discovery. In recent decades, a growing number of experimental physicists, accustomed to accelerator experiments, have been dedicating themselves to this type of research. One of the first groups that made the transition in the 1990s was the one in which I was working at CERN with Sam Ting, engaged in the study of the Z^0 boson. We began discussing how to address some of the unresolved issues in elementary particle physics by studying those cosmic signals from space. As we will discuss in the following chapter, the focus soon turned to the problem of the disappearance of primordial antimatter, currently one of cosmology's greatest mysteries.

26

A SPECIAL PARTICLE

We have seen how few ingredients are actually needed for building the universe. Among these, there is energy, which can take on various forms, including that of the mass of the various elementary particles, which can therefore be compared to containers in which a large quantity of energy is packed.

The problem is that the dimensions of these containers—the masses of the various particles—have very different values. Between the smallest masses, those of neutrinos, and the largest, that of the top quark, there are about fifteen orders of magnitude; that's approximately the mass ratio that exists between a sperm and a man. And that's not all: the mass values of the particles are fixed, but no one has yet understood what the regularity behind the observed values is.

In short, mass is one of the fundamental quantities, like space, time, and energy, that play a crucial role in defining the characteristics of the universe, but about which we have only a limited understanding. In the theoretical development that led to the definition of the Standard Model, which we talked about earlier, the question of mass has always been treated as a separate matter, a list of parameters determined by experimental observation. To be precise, the theoretical framework of the Standard Model was based on the (paradoxical) assumption that the elementary particles all had zero mass.

In 1960, Peter Higgs, an English theoretical physicist, made a major advance on the mass problem, associating mass with a property of the vacuum. This mechanism, which bears his name, predicted the existence of a new neutral particle, a boson with zero angular momentum (spin). The Higgs boson creates a field, like the electric or gravitational field, which has the property of producing the mass of the various particles. The more massive the particles are, the more the Higgs boson interacts with them. Yes, you got it right: the Higgs mechanism is like a glass that can be filled to different heights, where the heights corresponding to the different masses are observational parameters, not predicted by theory. These are parameters defined in the initial instants of the universe, when the properties of the vacuum crystallized, breaking the symmetry that characterized all particles and all interactions.

After the publication of Higgs' article, there was intense theoretical activity, followed by a systematic search for this particle at high-energy accelerators. The point was that the theory of Higgs and his colleagues did not determine the value of the boson's mass; therefore, it was necessary to search for it at all energy levels reachable by the most powerful particle colliders. I still remember discussions at CERN at the beginning of the 2000s, in which it was even questioned if the Higgs mechanism truly referred to a real particle. Toward the mid-1980s, CERN started developing the largest accelerator ever constructed, the LHC, with the goal of finding the Higgs boson. It was likely the most intensive effort ever made to reach a purely scientific goal. The energy available in the collisions between the protons in the two beams would reach the astonishing value of 16,000 times the proton's mass. The experimental apparatus developed to study the results of these collisions and look for the very rare cases of the production of a Higgs boson was as large as a four-story building and the experiments were carried out with the most sophisticated detection techniques and the most advanced electronics. The teams of scientists who made up the international collaborations involved in each of the various experiments included more than 5,000 participants, with observational activities that spanned decades and continue to this day. A huge effort was rewarded with success. The announcement of the discovery of the Higgs boson, a particle with a mass about 133 times that of a proton, was made at CERN on July 4, 2012. The Nobel Prize was awarded to Peter

Higgs and the Belgian theoretical physicist François Englert the following year.

Why was the discovery of the Higgs boson so important? Because it represents an essential confirmation of the theory of fundamental force fields and elementary particles, with deep implications regarding the structure of the vacuum and its stability, both at the moment of the Big Bang and today. To give just one example: from the value of the mass of the Higgs boson and the mass of the heaviest quark, the t quark, considering current experimental error, we derive that the vacuum of our universe is, potentially, a metastable vacuum. If this were to be confirmed by more precise measurements of the mass of these particles, it would follow that on timescales of billions of years, the physical vacuum could transition to a more stable state. The last time that happened was, as we saw at the beginning of the book, during the Big Bang! This shows us what an important step forward this discovery is for a field of physics that is on the frontier of knowledge. As has been said more than once: we know that we don't know and there are still many things to be clarified, but real science moves forward systematically, step by step, reconstructing the puzzle of the universe. Sometimes it has taken a century to add one piece, as with gravitational waves, in others fifty years, as was the case for the Higgs boson. It might even take millennia, like confirming the heliocentric theory of the solar system. But every time we reach a goal and manage to illuminate a larger part of the stage, the spectacle that presents itself justifies the effort made and pushes us to move forward.

27

ANTIMATTER

We have seen how the universe derives from a vacuum state: every possible type of conserved quantity, energy, electric charge, the number of leptons, the number of baryons, must equal zero, both at the beginning and in every step that follows. This elementary consideration led to the conclusion that matter and antimatter were also present in perfectly symmetrical quantities at the beginning of the universe. Today, however, there is only a negligible amount of antimatter in the universe, to the point that we wonder if, apart from that created every day by collisions between relativistic particles, the type of primordial antimatter produced during the Big Bang has disappeared completely. This question is one of the most fascinating problems in contemporary physics. In the description of the first instants of the universe, we mentioned *baryogenesis*, which is to say the disappearance of primordial antimatter in a great flash, in which almost all matter was transformed into energy. The tiny fraction of surviving matter—we are talking about parts per billion—could be the result of a mysterious asymmetry between matter and antimatter that, extrapolated to the present day, has led to a total asymmetry in favor of matter in what surrounds us—at least in our corner of the universe.

But what is antimatter? It's like trying to spot the difference between two images in a book of brain teasers. If we look around, we know that we are surrounded by atoms made up of electrons, protons, and neutrons, immersed in omnipresent electromagnetic radiation. All of this is matter. But we also know something else: that the innermost structure of matter is symmetrical and that its opposite, antimatter, exists. This means that every type of particle has its own antiparticle.

We owe the concept of antimatter to Paul Adrien Maurice Dirac, the enigmatic and profound English physicist, universally recognized as one of the fathers of quantum mechanics. In a letter to Paul Ehrenfest on August 23, 1926, Einstein wrote of him: "I have trouble with Dirac. This balancing on the dizzying path between genius and madness is awful." In 1928 Dirac wrote a very elegant equation that describes the fundamental properties of fermions, particles with a semi-integer spin like electrons, in a context consistent with quantum mechanics and special relativity. A genuine masterpiece, this immortal equation is engraved on his tombstone at Westminster. An equation too beautiful not to be true, but one that had a surprising consequence: the prediction of the existence of states (particles) with negative energy (mass). A particle with negative energy is not easy to accept. Among other things, such a particle would be able to endlessly lose positive energy, for example, by radiating photons, and reach lower and lower energy levels—truly strange behavior. On the other hand, the equation has many interesting aspects and correctly describes the properties of fermions.

To explain the solutions with negative energy, Dirac introduced the concept of the antiparticle, a fundamental state of matter characterized by having all quantum numbers assimilable to a charge (e.g., electric charge in electromagnetic interactions, color charge in the case of strong interactions) opposite those of the respective particle, with which it has the innate tendency to annihilate itself, producing energy in the form of photons. The latter, having the quantum numbers of the vacuum, can be produced at will, respecting the fundamental conservation laws.

Sometime later, Richard Feynman—using the diagrams that made him famous and which allow us to make what happens at the subatomic quantum scale visible, at least from a mathematical perspective—together with Ernst Stueckelberg, offered an interesting interpretation of the antiparticle state. An antiparticle is formally equivalent to a particle that is moving back in time.

In any case, Dirac got it right with his equation: antiparticles really do exist. In fact, in 1933 he was awarded the Nobel Prize, immediately after the discovery of the positron by Carl David Anderson, also recognized by the Royal Swedish Academy in 1936. The formulation of Dirac's equation and its consequences are one of those sensational cases in which abstract reasoning manages to anticipate new fundamental properties of nature. Since 1932, the idea of antimatter has attracted the attention of physicists, as has the fact that the annihilation between particles and antiparticles is compatible with the creation of energy, without altering the quantum numbers of the vacuum.

This property turned out to be very interesting when the Big Bang theory, developed by Lemaître with contributions by George Gamow and Ralph Alpher, began to be accepted in the late 1940s. Then came the problem of characterizing the initial phase of the universe from a physical perspective. Despite its complexity, it can always be broken down into a series of elementary interactions that respect the conservation laws of the fundamental quantities. So, if there was nothing before the Big Bang, we were therefore in a state compatible with the properties of the vacuum: even after the initial explosion, the total value of the quantities conserved in the fundamental interactions (energy, electric charge, baryon number, and other quantities) need not change. The existence of elementary antiparticles, which have exactly the same mass as the particles but opposite quantum numbers, perfectly respects these conditions, as long as the quantities of particles and antiparticles present in each instant after the Big Bang are the same.

That said, if there is something that immediately catches our attention, it is that we live in a corner of the universe, of which we don't know the size, dominated by matter. We know how to produce antimatter— even in the complex forms of relatively heavy antinuclei—by way of collisions between particles with relativistic energies, but, immersed in a world dominated by matter, the "life expectancy" of these antinuclei is minuscule. The observation of the universe gives us some indications, however preliminary, that suggest antimatter is present in negligible quantities, if at all, at every scale. This is a great mystery, given that at the beginning, antimatter—in the form of antiquarks and antileptons—was just as abundant as matter.

We should remember that the study of antimatter is carried out both here on Earth and in space. On the ground, antimatter is regularly produced in the laboratory, inside vacuum chambers, in controlled conditions in order to prevent it from interacting with matter and being annihilated. At CERN, macroscopic quantities of high-energy antimatter particles are being created, as was done for Rubbia and Van der Meer's research. It is also possible to create hydrogen antiatoms, one antiproton bound with a positron, and trap them within the appropriate type of magnetic fields in order to study their properties. So far, all of these atoms' properties have been found to be identical to their matter counterparts, with a precision level of one in 10 billion. The mystery of matter–antimatter asymmetry remains unsolved from an experimental perspective.

So, what can we do from space? Is it possible to determine if any antimatter structures exist by observing them from a distance? We are able to see and study the universe mainly due to the light emitted by different sources and at different frequencies; through radio waves emitted by super-cooled bodies, like gases and molecular agglomerates; and via gamma rays emitted by extremely violent nuclear reactions, passing through infrared, visible, and ultraviolet light and X-rays emitted by stars and bodies at high temperatures. However, since the properties of antimatter and matter are identical from an electromagnetic point of view, we will never be able to identify an antistar of a star by limiting ourselves to analyzing the light it emits. We are obliged, as Saint Thomas was, to "touch it," or better yet to search in cosmic rays for any traces of antinuclei, which might have been produced by nuclear fusion inside an antistar, that have managed to reach us.

Do you think that is a simple task? Well, it is something like catching a drop of ink mixed in with millions or billions of water drops falling in a rainstorm over New York. The data available to date does not allow us to make any categorical statements on the disappearance of antimatter; however, they do allow us to define what, in scientific terms, are known as *upper limits*, which is to say: *there is less than a fixed quantity.* Currently, that limit is less than one helium antinucleus for every 10 million helium nuclei present in cosmic rays. Conversely, it would be enough to observe some helium antinuclei with certainty to be able to establish the existence of relevant quantities of antimatter, given that only fusion reactions in antistars are capable of producing antinuclei, such as antihelium or heavier. It would be a momentous discovery, based on the detection of a few nuclei with a charge that is "wrong" compared with what is listed on Mendeleev's table.

This is this idea that, in 1993, motivated the development of the experiment to which I dedicated twenty-five years of my scientific efforts and which we will discuss in the following chapter: the Alpha Magnetic Spectrometer (AMS), the antimatter hunter on the International Space Station.

28

HUNTING FOR ANTIMATTER IN SPACE

Geneva, summer 1993. At CERN, in the office of Sam Ting, 1976 Nobel laureate for the discovery of the J/Psi particle in the famous November revolution, which was mentioned earlier. After graduation in the early 1980s, I left Sam's research group to participate in two different experiments, each important in its own way. With the first, UA2, I took part in the discovery of the intermediate bosons Z^0 and $W^\pm$, for which Carlo Rubbia and Simon van der Meer received the Nobel Prize in 1984. The second was the Stanford Large Detector (SLD), at the Stanford Linear Accelerator Center (SLAC) in the United States, where we were trying to produce large quantities of the newly discovered Z^0 bosons, by adapting the existing accelerator, aiming to reach the goal before the dedicated accelerator, the Large Electron–Positron Collider (LEP), which was being built at CERN in the meantime.

In the field of frontier research, deciding which direction to take is a bit like trying to pass on the curve in a Formula 1 race—it can go well or it can go badly. If the choice to participate in UA2 turned out to be a success due to the results, the same cannot be said for the experience at Stanford. Modifying the SLAC accelerator was slow and difficult; it was finally completed well after CERN's LEP was up and running in the late 1980s.

At the end of that decade, I went back to work with Sam Ting, who in the meantime had formed a large research group, the L3 Collaboration, and carried out the most impressive of the four experiments at the LEP. Sam suggested that I coordinate the construction of an innovative detector made with silicon. It was an ultra-precise instrument; I was completely occupied with its construction for the next four years. By 1993, Ting's LEP experiment had already entered the phase in which it would collect data without substantial modifications until the 2000s. So, Sam was already looking for new ideas and challenges. At that time he had dedicated his energies to the creation of a super research collaboration with the goal of developing a colossal experiment that would exploit the new super proton accelerator being built in the United States: the Superconducting Super Collider (SSC). But also in this situation, things did not go as planned; in the fall of 1993 the SSC project was canceled by the US Congress. Sam had also tried to propose a derivative experiment at CERN for the new Large Hadron Collider (LHC) accelerator, then under construction, with the aim of discovering the Higgs boson. However, CERN, under the direction of the new Nobel laureate Carlo Rubbia, rejected the proposal.

Particle physics experiments require lengthy planning. First, a scientific proposal has to be drafted and presented to funding agencies, then a collaboration has to be formed—that is, a group of scientists, often hundreds and sometimes thousands of researchers and engineers—that will be responsible for creating the sophisticated instruments that will be used in the experiment. The construction, calibration, and implementation of a particle accelerator experiment is a highly complex activity: you have to position and connect dozens of kilometers of cables and build rooms full of electronics for acquiring signals and full of computers for data processing. This is why you have to start working on an experiment five to seven years before the moment when the data starts to be collected and analyzed. It is as if Galileo, in order to study the motion of the pendulum in Pisa's cathedral, had needed to build the cathedral first: he wouldn't have been able to work alone and, in any case, he could have done only a few experiments in his lifetime. If we did not want to lose the treasure trove of skills gathered in the L3 Collaboration—made up of around 600 extremely capable researchers who had been working closely for decades—we had to make our move and identify an interesting, stimulating scientific prospect. In the summer of 1993, discussion in our working group had rapidly shifted from particle physics to astroparticle physics, and from ground-based physics to space-based physics.

Those were the years in which many of the fundamental research sectors that are well established today were just getting started. We considered moving forward in the field of high-energy gamma rays in space and on the ground, but we realized that working in astrophysics would have taken us away from fundamental physics. The search for gravitational waves with laser interferometers was still far from reaching meaningful goals, not to mention that it required extremely complex technology. In the end, we decided to focus on the search for antimatter in cosmic rays, an ambitious and fascinating goal that was perfectly in line with our field of research.

As mentioned in the previous chapter, the laws of nature, insofar as we know them, are symmetrical between matter and antimatter. At the same time, matter and antimatter cannot coexist without destroying each other. On Earth, the little antimatter produced in radioactive decay or in interactions between cosmic rays and the atmosphere is primarily composed of positrons, as well as a few antiprotons and unstable particles, such as positive muons and charged pions. At hospitals, positron emission tomography (PET) scans are done with antimatter, that is, the positrons emitted by the radioactive isotopes; this is the only way to produce molecules present in human physiology and create images of certain types of tumors.

In space, the question is fundamentally different. Elementary antimatter particles, possibly present in cosmic rays, can survive and travel for extensive distances. The average density of interstellar matter is very low, on the order of one hydrogen atom per cubic centimeter. We know that a small quantity of antiparticles, produced by random collisions with interstellar matter, exists in cosmic rays; we are talking about one part in 200 for positrons and one part in 10,000 for antiprotons. Antineutrons are unstable, like neutrons, and last for only a few minutes.

The new team's objective was to build a particle detector in space that could analyze the flow of cosmic rays, looking for ultra-rare components at a level of parts per 10 billion. The discovery of heavy antiparticles, like antihelium-4, would have been completely unexpected, as it would have required the existence of nucleosynthesis processes possible only within antistars.

The experiment was dubbed the Alpha Magnetic Spectrometer (AMS), referring to the space station that was still called Alpha at the time and had not yet been built. During numerous meetings at CERN and MIT, the details of the experiment were defined.

The idea was to use the same technologies developed for the accelerators and adapt them for space. In the search for antimatter, it is essential to measure electric charge: particles and antiparticles have opposite charges. In order to do this, it is necessary to observe the curvature of the path of every single cosmic ray that passes through the experiment, using a magnet and a track detector. It was a new challenge for all of us: to adapt an elementary particle detector we had always used on the ground for a shuttle launch and operation in space. Many aspects were as yet undefined so we left ourselves some alternatives. For example, two types of track detectors were hypothesized: one traditional, proposed by MIT and based on a large, gas-filled volume, capable of making a lot of position measurements, each one of limited accuracy; the other, proposed by my team at Perugia's Instituto Nazionale di Fisica Nucleare (INFN), based both on the developments done for the experiment at LEP and on a few plans for silicon-based detectors, each one capable of providing very high-precision position measurements. Following a long series of tests, it turned out that my proposal was significantly more suitable for use in space and was approved. It was the first large silicon detector to be operated in a non-Earth environment. Once completed, we submitted the proposal for the AMS experiment to NASA, INFN, ASI, and to other space agencies and research bodies in sixteen different countries. NASA administrator Daniel Goldin was quite pleased with the idea and he approved a test flight on the Shuttle to be carried out in 1998.

It was 1994 and we had fewer than four years to get ready!

It seemed an almost impossible feat for a team that, despite its impressive experience in particle detectors, had no experience in the space sector. Goldin provided us with support from an excellent team of engineers from NASA and Lockheed Martin, who were essential. It was four unforgettable years of intense work but also of unprecedented enthusiasm. We were entering a new world, in terms of both the scientific objectives and the technologies to be developed. On June 2, 1998, AMS-01, the test version of AMS, was launched with the STS-91 mission and operated successfully for nine days in orbit, gathering hundreds of millions of cosmic rays and verifying that the amount of antihelium was less than one part per million.

The success of AMS-01 allowed us to continue building the final version, AMS-02, which would be able to collect 10,000 times the statistics, operating for three years on the International Space Station. However, in 1998 we couldn't have known that 2003 would be the year of the Space Shuttle *Columbia* disaster, a tragic accident in which the spacecraft disintegrated during reentry, killing all seven astronauts on board. Consequently, the launch of AMS-02 would take place thirteen years later. It was a complicated time. In 2003, NASA removed the AMS-02 from the shuttle program at the request of the Columbia Accident Investigation Board in order to minimize the number of missions needed to complete the International Space Station. We found ourselves in a paradoxical situation: committed to building a detector that would never fly.

At that time I had become the Deputy Spokesperson for AMS-02 and we began, with Ting, systematically lobbying the US Congress to be reinstated. It was a highly instructive opportunity to see how the American political system works in practice: in meeting with the main Democratic and Republican senators and representatives who dealt with this issue, I was impressed by their competence and the attention and respect they had for us. For them, space, scientific research, and international collaboration were to be supported in a bipartisan way, without useless or specious ideological conflicts. The decision was made. NASA's new administrator, Michael Griffin, was instructed by Congress to reinstate AMS-02 in the shuttle program.

Unlike the initial plan, according to which the AMS-02 would have been returned to Earth every three years to be refueled with liquid helium—necessary to make the new superconducting magnet we were developing work—in the new plan it would be permanently installed on the International Space Station with no option for bringing it back. This led to the decision to reuse the permanent magnet successfully deployed on AMS-01. On May 19, 2011, the AMS-02 was launched as part of the Space Shuttle's penultimate mission: STS124. On board the spacecraft was an Italian astronaut, Roberto Vittori, who was to operate the robotic arm of the ISS during the delicate mounting of the AMS-02 on the main structure of the space station, where another Italian astronaut on his first long-term mission, Paolo Nespoli, was waiting for him. Eight years later, AMS-02 has collected almost 150 billion cosmic rays, studying the composition of this shower of particles hitting the Earth from the depths of the cosmos in unprecedented detail. A number of new effects related to the properties of cosmic rays have been found. As for antimatter, an unexpected excess of high-energy positrons has been discovered, peaking at an energy of around 400 times the mass of a proton, the source of which remains a mystery, for now. It could be a never-before-observed astrophysical phenomenon (positrons produced by pulsars) or even a sign linked to the indirect observation of dark matter particles of substantial mass. As for nuclear antimatter, eight antihelium candidate events have been recorded so far: six of antihelium-3 and two of antihelium-4. These are ultra-rare events, gathered at a rate of two antihelium events per 100 million helium nuclei.

As we discussed earlier, the indisputable observation of a few helium-4 antinuclei would have profound consequences on our understanding of the universe. That is why these events have not yet been published in a scientific journal, and are being studied very closely by the AMS-02 researchers in order to understand if there might be unknown instrumental background sources. As Carl Sagan said: "Extraordinary claims require extraordinary evidence." Stay tuned!

29

NEW TERRITORIES, NEW DAWNS

In the previous chapters we looked at the immensity of the cosmos, which extends immeasurably in time and space relative to humanity's brief history and even briefer individual experiences. Faced with this awareness, we can be surprised, filled with admiration, and intimidated. In any case, not knowing whether or not we are alone in the universe or the perception of our inevitable limitations certainly won't stop our desire for progress or diminish the fascination of the unknown. On the contrary, it is the perception of those very limits that is the most powerful driver of exploration. Just as Christopher Columbus embarked on the search for the shortest route to the Indies, even if based on erroneous assumptions, and landed in the Americas, so one of the most visionary challenges of our times is to reach and colonize other planets.

Speaking of visionaries, a great Russian engineer and scientist Konstantin Tsiolkovsky, an astronautics pioneer who lived in the second half of the nineteenth century, argued that "Earth is the cradle of humanity, but one cannot remain in the cradle forever." There is certainly some truth to this statement. The problems are time, technology, and opportunities. In 1957, exactly a century after Tsiolkovsky's birth, Sputnik was launched and humanity ventured into space for the first time, moving beyond the Earth's atmosphere. Then an unprecedented acceleration began; with the space race between the United States and the USSR and constant technological innovations, just twelve years after Sputnik, in 1969, a man walked on the Moon.

The Apollo program, started in 1961, lasted for eleven years, with six moon landings and one mission, Apollo 13, aborted after a catastrophic explosion near the Moon, from which the crew miraculously returned to Earth safe and sound. However, we did not stay on the Moon. After the Apollo program, the superpowers' ambitions shifted. We concentrated on building increasingly sophisticated space stations and laboratories located in low Earth orbit. Then came the construction of the International Space Station, operational since 2000, involving the collaboration of space agencies of the United States, Russia, Europe, Japan, and Canada.

In the meantime, other powers have made forays into space: India and, above all, China, which has independently developed an ambitious program, including crewed missions, for now on a space station in low orbit, one day on the Moon. As president of the Italian Space Agency, I participated in a number of meetings with the heads of space agencies around the world to discuss the prospects of human space exploration. Until 2017, we were working on a joint program that was focused on the exploration of Mars. Charles Bolden, an astronaut and NASA administrator for eight years, was a tireless supporter of a global program that would involve all of the major agencies, China and India included. Then Trump was elected and multilateralism in the space sector suffered a major setback. Stuck between the commercialization and militarization of space, NASA is now aiming for a return to the Moon and the launch of a cislunar space station to serve as a portal for robotic and human missions to our satellite. This complex program is still being defined and should be ongoing through the middle of this decade. In short, sixty years after Armstrong's historic step on the Moon, we will still be at the starting gates. Or thereabouts.

And the other planets? Let's start by saying that landing the first person on Mars can wait, if we look at space agency plans. This is because the strategy of international collaboration in space is strongly affected by the political situation and by relations between the superpowers involved. Unfortunately, these relations have changed and have degraded significantly over the last few years. Space multilateralism was axed during

the Trump administration in favor of easier-to-control, bilateral relationships among "like minded" countries; space geopolitics has gained a
lot of momentum, supported by military, strategic and economic arguments. Fortunately, we are beginning to see some glimmers of hope. In
fact, today there are other players who could take the lead: exceptionally
wealthy private entrepreneurs who are investing their fortunes, accumulated through the new economy's global businesses, in space.

Undoubtedly, Elon Musk stands out; the now legendary head of
SpaceX makes no secret of his goal: to die of old age on Mars! On a number of occasions I have met with Elon and his team at the company's
Hawthorne headquarters, located in the southern part of San Francisco,
not far from the international airport. Each time it's a unique experience
because this South African–born inventor and entrepreneur (with US
citizenship) always has something new brewing, an idea that the whole
world is talking about or will be in the following days or months. An
industrial facility houses all of the activities related to the construction of
the launchers: design, research and development, assembly. This is where
SpaceX's 6,000-plus employees work. Standing fifty meters tall, in front
of the main entrance, is the first stage of the historic Falcon 9 rocket,
which was successfully recovered for the first time on December 21, 2015
(we'll talk more about that later). Like Musk's other businesses, nothing at
SpaceX is left to chance: Elon's hand and mind are evident in the overall
vision as well as the smallest details.

Let's start with the structure of the industrial facility. It consists of a
single, enormous parallelepiped into which pieces of metal enter and
rockets come out, ready to be transported to one of the three launch
centers in the United States: the Kennedy Space Center or Cape Canav-
eral Air Force Base in Florida or Vandenberg Air Force Base in California.
This organizational peculiarity is also a significant competitive advantage
when compared with the fragmentation of similar industrial structures
used for the construction of launchers developed by NASA (Atlas, Shuttle,
SLS) or ESA (Ariane, Vega).

The reason for the difference is clear: a public space agency represent-
ing an entire continent is obliged to distribute manufacturing activities
throughout those territories that contribute to financing its programs. All
of this reduces efficiency and increases costs.

The organization of SpaceX is extremely functional: the front of the factory is a gigantic open space, housing hundreds of engineers and administrative staff. Elon's personal office is one of the many common spaces, just a little larger than the others. In the center are the meeting rooms. It's a large, insulated, silent area full of people, computers, and keyboards.

A very long wall separates this area from the production area, which is accessed in a decidedly cinematic way: through a door below one of the Falcon 9's colossal landing legs. Hanging from the ceiling, you can also admire one of the first versions of the Dragon capsule to successfully return to Earth, scorch marks included. To the right is the control room, from which the launch phases are followed, which became famous when Musk organized livestreaming of some of the events. An elevator takes you to the upper floors, past display windows showcasing mannequins in space suits from Hollywood films, together with the latest versions of the suits designed for the first SpaceX astronauts who will pilot the Dragon when it carries the NASA and ESA crews into space, along with some private citizens. The café-restaurant, open 24/7, is free. It has no walls and is adjacent to the assembly lines. There are no separations between one sector and another. The impression is of an anthill. The characteristic melting pot of American society can be seen in the faces of the hundreds of people who, dressed in casual clothes, work, talk, meet, and move in every direction. Every facet of this multicultural society is in evidence.

Incidentally, every time I visit SpaceX, I discover that something has changed about the way the work areas are arranged. For example, since they have succeeded in systematically recovering the first stage of the Falcon 9 by landing it again after launch, they have created a new area where the "recovered" Merlin liquid oxygen engines are inspected and refurbished, to prepare them for reuse. This activity started just two years ago and has now become an integral part of production. The same is true of reusing the first stage of the rocket, which requires careful checking of the structural integrity of the mechanical parts. For some time now, the spectacular images of the recovery of the Falcon 9 and the engines of the Falcon Heavy show that the main components of the launcher are stained and discolored. There is a reason for it. It's to show that they've been reused: there is no reason to repaint them after they've returned from space.

As for the various manufacturing processes, they range from the engines to the stage fuselages to the machines for 3D molding of engine parts to the molding system for the rocket nose cone, or *fairings*, formed of two shells that protect the payload as it penetrates the atmosphere. There are no noisy processes, just a low buzz generated by the ongoing activities. Yet no one is wasting time; everyone is following an invisible but functional plan.

As you get closer to the furthest reaches of the facility, the rockets begin to take shape: enormous horizontal cylinders, one after another. Like what you would see in an automobile factory but with two substantial differences: here there is almost no automation—most of the work is being done by hand, entrusted to hundreds of technicians and engineers. The second is that the stages of the Falcon 9 are 20–50 meters long and are completed in about a week. I meet with the head of production and planning: he is an Englishman who worked in the European automotive industry, on production lines capable of turning out hundreds of thousands of vehicles per year. He explains how he is handling the production of thirty Falcon 9s and mentions, lowering his voice and with eyes sparkling, that he could easily increase production by a factor of ten.

But this might never happen—not for lack of customers, but because of the SpaceX revolution in the launcher sector we just mentioned. Up until December 21, 2015, the launchers were used just as they were by the Chinese, who invented solid fuel rockets in the thirteenth century for military use as well as for launching fascinating fireworks. The rocket was basically a tube filled with fuel that was destroyed as it was used. And that's not all. The violence of the chemical reaction of solid fuel, necessary for propulsion, also tends to damage the casing, making it unusable even if recovered. Furthermore, the reentry of the stages from the higher atmospheric zones usually ended catastrophically: their considerable weight makes it impractical to resort to parachutes. The advent of liquid propulsion—conceived around 1930 in the United States and in Germany, but only put into production on an industrial scale by the Nazis with Wernher von Braun's V2 rockets—paved the way for a different solution.

First of all, liquid propulsion takes place inside a special combustion chamber in which the two components react, without involving the tank as in the case of solid fuel. This is a huge advantage, because it means that the structural components of the rocket's first stage are not damaged during liftoff. What's more, in the same way the propulsion can push the rocket to speeds high enough to supply the satellite with the velocity it needs to stay in orbit, the same engines can also brake the rocket's fall and bring it back to earth without damaging it.

In principle.

The truth is, no one really believed it. It seemed impossible, but no one had even seriously tried; in fact, rockets continued to be built to be discarded as soon as they were used. Utter madness—it would be like building a new airliner every time we crossed the Atlantic. And, at a cost considerably more than $500 per person, the ticket price would be 1,000 times higher.

Then Elon Musk decided to try first-stage rocket recovery. Within a couple of years and after a few sensational failures, cleverly announced and fearlessly broadcast worldwide, on December 21, 2015, the first stage of a Falcon 9 successfully landed on a floating launch platform for the first time. This is an important date and should also be remembered for another reason. In Luxembourg, an important Ministerial Council of the European Space Agency (ESA) was held in December 2014. It marked a substantial change in the way Europeans approached the construction and marketing of launchers. After long and complicated discussions, it was decided that it was necessary to upgrade the family of European launchers: the Ariane 5 and the Vega. At the same time, investment by the ESA countries would have been limited to new infrastructure, while the marketing management for launch services would be awarded to private industry. This was revolutionary, a decision made precisely to confront aggressive competition from SpaceX and new private players in the industry. As the president of the ASI, I took part in many of these preparatory discussions and remember clearly that none of the French, Italian, or German experts acknowledged the importance of SpaceX's ongoing attempts to recover the first stage of their rocket. That is why the success of Falcon 9, just a year after the decisions made in Luxembourg, came as an unpleasant shock to the European space community. Today, this technology is well established and the recovery of the first stage takes place on a regular basis, both on land and on floating launch platforms. All of this also has significant consequences on industrial activities: not only are the prices of access to space lowered, but the number of new launchers produced every year is reduced! It's probably no coincidence that, at the end of 2018 and at the height of its success, SpaceX fired 10 percent of its workforce, around 600 people.

As we said at the beginning of the chapter, SpaceX is just the first step in Musk's long-term strategy. The goal is much more ambitious: to colonize Mars.

To face this type of challenge, in addition to extraordinary determination and financial means, a number of technological tools are needed, of which the launcher is only the first, albeit fundamental, ingredient. Several years ago, Elon had already started production of the Falcon Heavy, a launcher with twenty-seven engines consisting of three Falcon 9 first stages. The Falcon 9 could carry 4 tons of cargo to Mars; the Falcon Heavy could carry 16 tons. To date, two Falcon Heavy vehicles have been successfully launched and five of the six first stages have been successfully recovered. But, since he demonstrated the recoverability of the stages, Elon once again revolutionized his strategy. Because he wants to go to Mars with a group of "colonists," along with all of the equipment to build the necessary infrastructure, an even more powerful launcher is required. A launcher capable of carrying 100 tons to the red planet: the Big Falcon Rocket (BFR), now called Starship, powered by thirty-one powerful new Raptor engines.

It must be said that conventional techniques don't make a traditional approach feasible. In fact, it takes an incredible amount of fuel to leave the Earth. Let's remember that, put together, the command capsule and the Apollo 11 lunar module weighed 45 tons, while the Saturn V launcher weighed 3,000 tons.

Musk is looking at a very different scenario. Once Starship reaches low Earth orbit, it will stay parked there and be refueled by other service rockets that come and go from the ground. Once the fuel tanks are "topped up," it will be possible to transport a great deal more weight and significantly larger volumes to Mars than the Apollo 11 mission took to the Moon. Not only that, once you near the red planet, the landing will be implemented with the same braking technique developed with the Falcon 9. The procedure is the same for the return, this time having synthesized the fuel on the Martian surface, starting with CO_2 from the atmosphere and hydrogen obtained from underground water ice. The SpaceX team is engaged in this adventure, like the crews on Columbus's ships; the difference is that the SpaceX engineers freely participated in the overall project and they are highly motivated. In order to challenge the limits of technology and human capacity, one must have trust in progress, perhaps even faith, but once you start considering ideas at that level, many things become normal.

On the other hand, building a colony on Mars is a complex task. You have to find the right landing site, an area where the underground ice is available and easily extracted. Considerable amounts of energy are needed to synthesize fuel from the atmosphere and water, creating methane and oxygen—it is by no means clear how this energy is to be obtained. On Mars the solar constant is about 40 percent lower than it is on Earth. As if that weren't enough, sandstorms occur quite frequently, and in some cases they cover the entire planet for weeks or even months, taking that percentage to zero. The option of using nuclear reactors may be the only way to solve these kinds of problems.

In order to host a human colony, some infrastructure would also have to be built to protect the colonists from solar and cosmic radiation; otherwise, no one will be able to survive for long. The problem is that, at the moment, we do not have the intelligent robotics available to carry out these tasks without human help. Even the amazing NASA-JPL Martian robots—Spirit, Curiosity, Opportunity—are inadequate for managing the kind of construction site Musk has in mind, if for no other reason than their slowness. Since the Opportunity rover landed on Mars fourteen years ago, it has covered about 50 kilometers—a little more than 3 kilometers per year! We are confronted with the paradox of the chicken and the egg. For a colony to settle, we need robots that only humans can properly command, ergo we need a Martian colony! Problems, problems, problems. At the same time, by participating in the closed-door meetings held at SpaceX—to which experts from NASA and other international space agencies were invited, along with representatives from private industries and planetologists—it would be useless to deny that one learns and understands a lot of things. For example, I understand why the US company Caterpillar is often there; among other things, they deal with construction machinery. Today, most open-pit mines are excavated by robots controlled by a small number of operators. So, the robotic jobsite already exists on Earth, but it is based on human–machine interaction that presupposes physical proximity. In short, a crewed station could be built and put into orbit around Mars from which the robots working on the surface could be controlled at a distance. Thus the machines would have built the village before the colonists arrived to take possession of it.

Finally, there is the issue of ice, the raw material without which it is not possible to synthesize fuel or guarantee the survival of the colony. It must be said that, when it comes to water on Mars, Italians are the world's top experts. In 2003 an ESA spacecraft, *Mars Express*, was put into orbit around the red planet carrying the MARSIS radar experiment, equipped with an extraordinary triple antenna, made up of one 7-meter and two 20-meter segments. MARSIS is just one of the masterpieces created by the late Professor Giovanni Picardi and his team at Rome's La Sapienza University. The ground-penetrating radiography made possible by this antenna enabled the search for liquid and solid layers in the subsurface. In 2018 subterranean lakes were discovered about 1.5 kilometers below the surface, in the area of the south pole. But because the sounder was also sensitive to solid water, MARSIS allowed the mapping of extensive areas of the surface, highlighting the sinuous ridges characteristic of glacial movement, in this case covered with a light layer of soil that prevents its sublimation in the thin Martian atmosphere. So, it will be up to this group of Italian scientists to help define the best landing sites for Starship.

When I speak with the SpaceX engineers, I always come away with the same impression: they know the technology inside and out, to the point that they will reinvent the form and function of a screw or a bolt if that's what it takes to optimize a new application. They are not afraid of making mistakes; if they're not able to pull it off, they get down to brass tacks and work as hard as it takes. The dynamics of the professional relationships are clear and at the same time tough, competitive. In that context, in which you're working toward concrete objectives, you can make or break a team very quickly. All this, of course, is also a direct consequence of the enormous size of the American labor market in the space and technology sectors. If new business opportunities arise, Americans can recruit the competent resources necessary in their own backyard.

I got to know Tom Mueller, also known as the "first employee" of SpaceX. Originally from Idaho, Tom is the son of a logger and it seemed he would follow in his father's footsteps. He was passionate about rockets from an early age and started by playing with models. His innate curiosity led him to increasingly elaborate experiments, to the point that, using his father's tools, he turned an oxyacetylene welder into a rocket engine and modified the mixture to get a fuel that would produce greater propulsive thrust. He paid for his university education by working as a logger in the summers and obtained his engineering degree, finally establishing a career as an engineer and developer of liquid-fuel engines. Elon often says that Tom is one of the best investments he's ever made. Over a cup of coffee, Tom told me about how he was working on the development of the BFR's new Raptor engine, which in August 2019 successfully completed its first test flight with Starhopper in the Texas desert.

Another notable person on the SpaceX team is Gwynne Shotwell, the current president of SpaceX. The eleventh SpaceX employee, she started working with Elon in 2002. With a background in science, math, and engineering, she works at the top of the space industry and is currently one of the world's most influential women, engaging, direct, and visionary. I remember a discussion with her and some members of her team in front of an incongruous gas fireplace, blazing away despite the fact that it was summer in California. We discussed an exceptionally broad range of topics. The dangers of cosmic radiation for astronauts traveling to Mars. How to protect them during interplanetary travel (with large superconducting magnets) but also later, once they get to Mars. Which scientific instruments should be taken to the red planet and for what purpose. About quantum mechanics and its applications. All that in addition to, naturally, the increasingly powerful rockets capable of going into orbit and exploring the cosmos.

In Europe, I have found myself participating in this type of discussion only in the context of universities or in the research world. These topics have rarely been broached in talks with the top management of companies and industries, who are too busy managing the economic and political aspects of their activities.

So, at the public level and beyond, we ask ourselves with increasing urgency: How do we bridge the sometimes excessive gap between industry and research? I think that all we have to do is observe the role engineers play in the management of innovative start-ups and take into account the technical depth that an industrial manager must demonstrate to be credible in such contexts. That's not to say that in the United States one doesn't find companies in which lawyers play a more important role than engineers, but the American ecosystem, the way people work and the ability to take risks, manages to support a level of innovation that is capable of challenging the established players.

This is where, in Europe, and in Italy, we need to seriously reevaluate how we do things and implement some drastic measures. As the SpaceX story demonstrates, in the current climate, an industry that manufactures large rockets for the civilian market can be revolutionized in 10 years. In other words, space has the power to inspire not only the dreams of children, but those of adults as well. Elon Musk is in good company: Jeff Bezos of Amazon, Paul Allen of Microsoft (just recently deceased), as well as Richard Branson of Virgin and Yuri Milner of Breakthrough, are among the most famous examples of entrepreneurs who built their success in industries other than space, but who then invested in the space adventure. Bezos founded Blue Origin in 2000; the company has developed a recoverable rocket and capsule that, in 2015, successfully passed what is known as the Kármán line, at an altitude of 100 kilometers, which conventionally separates the upper atmosphere from space. Subsequently, Bezos initiated the development of another large rocket, the New Glenn, which is also recoverable and reusable. Fascinated with space since childhood, Bezos's goal is to create transport systems that will allow future generations to aim for other destinations in the solar system: the Moon, Mars, and the frozen satellites of Jupiter and Saturn, where the presence of water offers tantalizing possibilities. In order to achieve this, he is investing a billion dollars per year from the sale of a small portion of his shares in Amazon.

I will talk about Milner later. But if we want to understand Allen's and Branson's dreams, we have to go to the Mojave Desert.

30

THE DAWN OF ASTRONAUTICS

If there is a place to break records, that place would be Space Port in California's Mojave Desert, a legendary location for aeronautics. When I visited it, it was like stepping into a movie. On the walls of the restaurant at the small airport is the story of the most recent records in aviation history; it's the terrestrial version of the intergalactic bar in *Star Wars*, minus the monsters, but filled with no less extraordinary characters. It is here, in the hangers surrounding the runway, that some of the most innovative aeronautical initiatives were born and developed.

One example is Scaled Composites, the company founded in 1982 by Burt Rutan, one of history's most brilliant aeronautical engineers and designer of forty-six different airplanes, five of which are exhibited at the National Air and Space Museum in Washington. He designed the Model 76 Voyager, the first plane to fly nonstop around the world in nine days, piloted by his brother Dick in 1986. Rutan himself then managed to win the Ansari X Prize in 2004, building, with the financial support of Microsoft's Paul Allen, SpaceShipOne, an innovative variable geometry space plane, powered by a solid fuel rocket engine, which on September 29 and October 4 of 2004 crossed the 100-kilometer barrier, exceeding, at the point of maximum speed, Mach 3, about 3,600 kilometers per hour. Then, in 2011, Paul Allen and Burt Rutan founded Stratolaunch Systems in order to make the Stratobus—the largest aircraft in history—capable of launching, from an altitude of 15 kilometers, other launchers of considerable size, in turn able to carry up to 6 tons into low orbit. The gigantic hangar that houses the Stratobus is there, in Mojave, at the airport from which the plane completed its first flight on April 13, 2019. In fact, with the recent passing of its main financier, Paul Allen, the company temporarily stopped the development of this aircraft. Also in Mojave, you'll find the headquarters of Virgin Galactic, the company founded by English entrepreneur Richard Branson. Having developed numerous successful initiatives under the Virgin brand, ranging from records to airliners to gyms, he got into the space industry in 2011. His goal was to launch, for the first time in human history, a service focused on "space tourism." According to US Air Force and Navy rules, only those who have exceeded an altitude of 80 kilometers have the right to be called astronauts. To make this designation available to a broader public, beyond the restricted group of actual astronauts, Branson bought the rights to use the technology of SpaceShipOne with the aim of developing a larger version, SpaceShipTwo, along with a commercial service that would take six passengers to an altitude above 80 kilometers for a handful of minutes, at a cost of $250,000 each. As one might imagine, the initiative was met with great fanfare and the first 600 tickets were sold in the blink of an eye. It must also be said that the technological and organizational aspects have taken much longer than expected, due in part to an accident in 2014, in which a test pilot died. However, in February 2019, in addition to the two pilots, SpaceShipTwo transported one passenger into orbit for the first time, coming closer to the stage of commercial exploitation that will be carried out in another area of the United States, New Mexico, where Branson has already built Spaceport America, a futuristic-looking center where future tourist-astronauts will be trained. In fact, the first commercial flight took place on July 12, 2021. When I visited Virgin Galactic's facility in Mojave, I was able to relive, first in the simulator and then in flight, the exciting experience of the SpaceShipTwo suborbital flight. At the end of the visit, one of the Virgin team's pilots, the Italian Nicola Pecile—a former Italian Air Force pilot—took me on a flight over the Spaceport area in a single-engine aircraft, so that I could experience the particular type of landing that characterizes the final phase of the reentry of SpaceShipTwo. During the landing, the spaceplane can no longer use the rocket engine and is therefore forced to land like the Space Shuttle did, taking advantage of the lift of the wings and the braking effect of air friction (drag).

As we were flying over Mojave, I realized that this area of the desert is not so deserted after all. It's the ideal place for activities that require extreme and stable climate conditions. In addition to the Spaceport, Edwards Air Force Base isn't far away, and there are other, smaller airports, including some used only for the temporary or permanent parking of hundreds and hundreds of aircraft. Enormous photovoltaic systems shine, reflecting the sunlight, while in the distance we see the tracks used for setting speed records, which exploit the absolute flatness of the salty desert plains.

At this point one wonders if the advent of space tourism with Virgin Galactic's suborbital flights really represents the beginning of a new era in commercial aviation after all. Will it soon be possible to connect two cities on our planet in a very short time? Will what is known as point-to-point suborbital flight be achieved, moving at high speed in space rather than at low speed through the atmosphere, thus drastically reducing flight duration? It's not that simple. To perform a suborbital flight that usefully connects cities located 10,000 kilometers apart, the velocity that must be attained in the phase of leaving the atmosphere is similar to that necessary for launching satellites into orbit. That means reaching a speed of about Mach 22, which only the thrust of rocket engine can provide. On leaving the atmosphere, SpaceShipOne and Two get only up to Mach 3, a velocity seven times lower and therefore a kinetic energy fifty times lower. This is the "limit" that, paradoxically, allowed Rutan to design a spaceplane with an elegant, slender, aerodynamic profile—similar to a typical airplane. In this case, the dissipation of energy during the reentry stage is limited and does not require special heat-shielding insulation on the surfaces. Despite this, in the ascent and reentry stages, the mechanical forces on the wings of the spaceplane are so high that changes had to be made to the aerodynamic form of SpaceShipOne—a decidedly brilliant and innovative solution. Unfortunately, it does not function at much higher speeds.

Where is the problem? Once you've left the atmosphere at high speed, you also have to worry about how to reduce this speed in the reentry phase. In short, you need to dissipate the accumulated energy, or, put simply, you have to brake. In flight, this braking can be done in only one of two ways: by activating a rocket that provides thrust in the opposite direction to the motion, or by using air friction. The first scenario requires a significant quantity of fuel, in addition to a special preliminary maneuver, which is a kind of somersault allowing the rocket engine to operate in the right direction.

If atmospheric friction is used instead, as it is for the reentry of crewed capsules, it would be necessary to protect the space plane from the enormous increase in surface temperature. Generally, special materials are used that, however, have their downsides: they weigh down the structure and consequently increase the power of the rocket engine necessary for the ascent. It's a bit like a dog chasing its own tail. Just think about the Space Shuttle; to put it into orbit we relied on fuel tanks and engines that were eighty times heavier than the shuttle itself. We all remember that the shape of the shuttle that reentered the Earth's atmosphere was very squat, due to the sophisticated thermal protection system that covered those parts exposed to friction.

Let's say that at the moment there is not a solution to this problem, with the exception of a rough idea inspired by Musk's Starship, the rocket concept for the colonization of Mars that we discussed in the previous pages: a powerful, recoverable launcher with liquid-fuel engines, capable of reaching Mach 20 with a few dozen passengers on board. According to this plan, after having traversed the necessary distance in space, the Starship flips over and brakes by restarting the engines before reentering the atmosphere at much lower speeds, then landing vertically, continuing to brake by using its re-ignitable engines.

So, we're talking about not a space plane but rather a completely recoverable and reusable launcher, a new type of aerospace transport system—potentially a decisive step for the birth of commercial astronautics, on this or other planets.

31

THE DAWN OF THE NEXT STAR

There are many stars, millions of millions, as an advertising jingle from the 1960s said. Today we know that in our galaxy alone there are hundreds of billions and that there are just as many visible galaxies. The Earth's closest neighboring star is Proxima Centauri, a red dwarf, one of the three stars in the constellation Alpha Centauri, about 4.4 light-years from us. Let's try to understand what such a distance means. We'll start with the solar system as a point of reference: its size is equal to about 15 light-hours. The Voyager 1 space probe, launched by NASA forty years ago, reached the boundary zone, called the heliopause, after about thirty-five years. A simple ratio allows us to calculate that, at this speed, it would take about 90,000 years to reach Alpha Centauri. Even the patient and heroic tardigrades we talked about would have a hard time dealing with such a long journey!

So, are we destined to remain confined to our own solar system? Let's say it is a question of speed: to go a long way in a reasonably short time, we need to move quickly. In the last century the speed at which human-made machines can travel has increased by a factor of about 1,000, from the legendary Marquise—the first production car made at the end of the nineteenth century, which could travel at 61 kilometers per hour—to the Voyager probe, launched in 1977, at 17 kilometers per second. How much more could the speed of a spacecraft increase? In reality by quite a bit, about a factor of 20,000, before reaching the impassable limit of the speed of light. Will we be able to increase it by a factor of 1,000 over the next century? What kind of propulsion could help us reach the stars?

The rockets we are used to are based on the principle of action and reaction created by violent chemical combustion reactions. The flame, violently ejected from the rocket nozzle, pushes the rest in the opposite direction. However, with this type of system, it's not possible to exceed Voyager's speed. A different method is needed, one that allows the exhaust gases to be expelled faster. This can be done, for example, with engines in which an electric field accelerates a beam of ionized atoms. But it takes a lot of power to drive them, and this limits the thrust that can be obtained. Increasing the available energy on board the rocket requires fuel that can produce a lot of energy, such as a nuclear reactor. However, these types of rockets are complicated to make, so much so that they have not yet made it out of the prototyping phase.

The truth is that we have to completely change our approach. The only thing that can accelerate something to a speed close to the speed of light is light itself! Using intense laser light beams, a small satellite attached to a solar sail can be pushed at high speeds. Unfortunately, the efficiency with which the light transfers its thrust to the solar sail is very low. If you want a microscopic satellite with a single gram of mass to reach a velocity of around one-quarter the speed of light, you need, even for a few minutes, a beam of laser light with 100 gigawatts of power, enough to power a quarter of Europe.

So, the challenge is very difficult, even if it is technically feasible. Perhaps one day picosatellites capable of reaching Alpha Centauri could be sent into space. It's such a fascinating idea that Yuri Milner, a billionaire financier of Israeli-Russian origins with a degree in theoretical physics from the Sakharov school, decided to finance a $100 million project, bringing together a crack team of scientists and engineers.

In 2012 Milner, together with his wife Julia, created the Breakthrough Prize, an initiative in which they were joined the following year by two other famous couples: Sergey Brin and Anne Wojcicki (Google and 23andMe), in addition to Mark Zuckerberg and Priscilla Chan (Facebook and the Chan Zuckerberg Initiative). This ambitious award aims to vie with the Nobel Prize in physics, biology, and mathematics (the latter is an area in which there is no Nobel Prize). The amount of the prize given is worth three Nobels and is awarded according to different rules from those established by the Stockholm Academy. For instance, it could be divided among all the thousands of members of the scientific collaborations who contributed to the discovery of the Higgs boson or gravitational waves. Between 2012 and 2017, the Breakthrough Prize gave out approximately $180 million; the awards are presented during a prestigious gala in Los Angeles, attended by Hollywood stars and members of the American jet set—decidedly more dramatic than the quiet, measured sobriety of the similar ceremony in Sweden.

The Breakthrough Foundation also manages Milner-funded scientific activities, including Starshot, the project to launch picoprobes toward Alpha Centauri. The goal is to have the picoprobes travel for around twelve to fourteen years to reach the star system, take a photograph, and send it back to Earth, where it would be collected four years later. I have met Milner a few times in his Silicon Valley mansion at the annual meetings dedicated to various facets of research related to life in the universe. These meetings are opened by the Starshot project director, Pete Worden, former director of NASA's Ames laboratory, with the message that any proposal or idea can be considered as long as it doesn't violate the laws of physics. These are thrilling meetings, where smart people tackle a variety of difficult problems with absolutely extraordinary scientific creativity. In one meeting we can discuss how to create a solar sail able to resist the intense beam of light coming from the ground and directed toward Alpha Centauri; in another, how to build a picosatellite weighing one gram, using a few square millimeters of silicon, covered with integrated circuits made directly on its surface. Or solutions are proposed for the almost certain possibility that such a long voyage will result in enormous radiation damage. In these meetings, time passes without anyone noticing. When you give a group of good scientists a complex problem, they will give their best.

Projects like Starshot are genuinely exciting because they show that we are already working on finding a way out of our solar system in order to become not only an *interplanetary* species, as Elon Musk would like, but also an *interstellar* species. Of course we need to get used to a completely different idea of travel; as I mentioned earlier, if we want to travel at the speed of light, we will have to send very small spacecraft, but doesn't all this remind us of something? Is it not likely that nature does the same thing to sow life between one star system and another? In fact, it is easier to send information out into space than mass. Who knows where this type of research will lead us? In any case, it's fun to think of a picosatellite that perhaps, one day, will arrive at Alpha Centauri, take a selfie, and send it back to those of us left on Earth.

Who knows what planetary sunrises are like in a three-star system?

32

EVEN MORE SATELLITES, EVEN SMALLER

Are we at the dawn of a new stage in space exploration? In what direction will technology develop? What opportunities will the future bring? We have already talked about the significant inroads made by private enterprise in an industry that, up until a few years ago, was reserved for governments and institutions. The contemporary space heroes—Musk, Bezos, Milner, Branson, and Allen—had never been to space before; they are entrepreneurs with technical skills and substantial personal and industrial assets. Leonardo da Vinci needed patrons to create new machines and amazing paintings; Galileo a university that gave him a professorship so that he could study the cosmos with his telescope; the same is true for Einstein with his revolutionary theories, even if at the beginning he got by through working in the Swiss patent office. Von Braun needed NASA to develop the Saturn V, after having developed the V2 in Nazi Germany. Today, the protagonists of the global space adventure are rarely astronauts; they are now figures who did not exist in the past. Their main characteristics are competence, vision, leadership, and, above all, the ability to control the flow of immense economic resources.

Technological revolutions can pave the way for the future if they are in the hands of the right people. The satellite miniaturization revolution is an exceptionally recent example, and the story of Planet Labs lays it out very effectively. Let's take a look at it. In 2010 three NASA engineers,

Chris Boshuizen, Will Marshall, and Robbie Schindler, founded a start-up in a California garage, and it was quickly dubbed Planet Labs. The idea was to build small, compact, low-cost satellites, called CubeSats, to create constellations for Earth observation (by constellations we mean a more or less numerous group of satellites, used in a coordinated and synchronized way). If we think about it, the idea made perfect sense given that, over the last two decades, technology has made great strides in both miniaturization and reducing the energy consumption of electronic components. The cell phone that we all carry in our pockets has many characteristics in common with a satellite, concentrated in an exceptionally small volume: it can communicate, geolocate its position, has an onboard computer, can gather and analyze images, and has a rechargeable battery. Equipped with a stabilization system and solar panels to recharge the batteries, your cell phone could do many of the things satellites do. Of course we have to be clear: while it's true that a single CubeSat cannot compete in terms of measurement power and accuracy, a large satellite cannot be everywhere at once, and, as we shall see, the effect of having many eyes working collaboratively can in some cases be an insurmountable advantage.

In 2013 the first four prototypes, called Doves, were put into orbit, and just a few months later Planet Labs announced a constellation of twenty-eight elements, launched from the International Space Station in February 2014. Planet's success was lightning-fast. Savvy investors immediately mobilized, and, catching a whiff of a great deal, they invested over $183 million in venture capital over two years. By the end of 2018, Planet had launched more than 300 satellites, went public, and became one of the two first "unicorns" of the new space economy (the other being SpaceX) (a unicorn is a private start-up that has reached a market value of $1 billion).

But what is the actual value of Planet's satellites? First of all, the spatial resolution (i.e., the capacity to record image details) is not exceptional: 3–5 meters. However, the most significant element of these constellations is that they are able to provide an image of every point on Earth once per day. A large satellite, which observes the Earth at high resolution, passes over the same point every five to twelve days, depending on its orbit. If it is specifically instructed to point to a given area, it can even review it once per day, but to the detriment of the rest of the observable surface. What is known as *revisit time* is a very important factor in monitoring human and natural activities that evolve at a daily rate, hence the considerable value of these images for agricultural, civil, marine, geological, infrastructure, and economic applications. Planet's images have radically changed the way we look at the Earth, given that a high frequency of revisits reveals effects never seen before. For example, observing the atolls in the South China Sea, we clearly see the progress of the fortifications and ports that China is building to control that part of the ocean. But—and here's the surprise—for the first time we saw that Vietnam is also discreetly working on some small atolls.

The exodus of refugees in remote areas of Africa can also be monitored, along with the daily trend of deforestation in unreachable areas of the Amazon and the daily status of water, gas, and hydrocarbon reserves of entire nations. Space data therefore provides a notable enrichment of terrestrial databases, allowing sophisticated analysis using big data techniques, highlighting phenomena that would otherwise go unnoticed.

The revolution is also just beginning. Today the number of nanosatellites launched into space is several hundred per year. From a strictly economic point of view, it must be said that building a constellation of a hundred satellites doesn't cost much; the most expensive part is the launch. They can even be replaced every few years, considering that, quite quickly, they can reenter the atmosphere and are destroyed. It should be emphasized that these instruments are not only useful in the field of Earth observation. Small constellations of nanosatellites can be extremely useful in a broad range of applications, for example, the distribution of cryptographic keys based on quantum telecommunications, an area that is certainly of interest to banks. Or in the realm of the Internet of Things: imagine a company that develops and sells its product around the globe, and that product requires maintenance, repair, resupply, and so on. A small constellation could gather this data in such a way that it helps to optimize logistics, reducing costs. This technology could also turn out to be very useful in the scientific field, as many aspects of our planet— such as the dynamics of the interactions between the solar wind and the Earth's magnetic field—would benefit from simultaneous measurements carried out from different points in space.

This sector is rapidly growing and evolving, so much so that mega-constellations of thousands of satellites are already under construction. OneWeb is a project of about 700 satellites designed to make the internet available everywhere in the world. Musk started the Starlink project, which plans to launch more than 12,000 satellites (the first nearly 1,700 are in orbit as of December 2021) to do the same thing but with even wider bandwidth. Undoubtedly, Musk has an advantage in that the SpaceX launchers are at his disposal and therefore he can considerably reduce the launch costs. It is interesting to note that in the past, particularly in the late 1990s, satellite constellations were already being considered as a possible solution to the telecommunications problem. Major investments were made in the certainty that cellphones would become satellite phones. Today we can see that this assessment was wrong. The satellite networks that developed were primarily land-based, with the cost of mobile phones gradually becoming lower and lower while the devices have become increasingly powerful. Twenty years later, the space industry is trying again. Inevitably, the objectives, costs, and market strategies have changed, but above all, the technologies have changed. If, on the one hand, it will be interesting to see the results of these new, impressive constellations, on the other, we need to carefully evaluate the impact of any collateral effects, which are by no means negligible. The massive implementation of this technology will increase the number of satellites operating in circumterrestrial space tenfold and, consequently, the number of potentially dangerous objects in orbit.

Getting back to the small satellites, it must be said that they can be used in a much broader range of applications, even extraterrestrial ones. A variety of scientific projects involving lunar or planetary exploration are being studied. These are based on nanosatellites launched directly from Earth or transported along with larger satellites and then put into interplanetary orbit at a specific time. A project developed by ASI, in collaboration with NASA, involves visiting a small, double-asteroid system with one asteroid in orbit around the other, Dydimos and Dydimoon; while the main NASA satellite will impact Dydimoon, the Italian nanosatellite will collect the impact images and send them back to Earth for scientific analysis. In anticipation of the exploration of Mars' or Jupiter's moons, the small satellites will prove ever more versatile and convenient tools to accompany, and eventually replace, the traditional satellites used for these missions.

But what has triggered the race for small satellites in the last ten years? After all, there have been no specific technological advances to justify this sudden interest. The truth is that certain things are created and develop without specific reasons. An opportunity simply presents itself, and someone, with the right skill set and good timing, takes advantage of it. A personal recollection in this regard might be useful. I became interested in small satellites back before they were big news, at the beginning of the 2000s, winning a series of university grants by presenting projects focused on creating a space qualification laboratory, built at the Terni location of the University of Perugia. Thanks to this financial support, the largest national university laboratory dedicated to the development of scientific instrumentation for use in space was completed in 2005. It was later exploited for the AMS experiment in direct collaboration with NASA (see chapter 28). I had been coordinating a group of researchers from INFN and various Italian universities involved in space exploration for ten years, and we were all energized and thrilled to be able to grow further in this sector, and become independent from the world of big industry. Part of the laboratory was dedicated to producing small satellites, a concept still largely undefined at that time, which people were beginning to see as potentially interesting. We made a series of proposals to the Italian space agency (ASI), which from 2005 to 2013 was plagued by particularly inefficient management when it came to financing innovation and scientific research. None of these proposals was accepted. We were able to obtain some minor, regional funding, but nothing came of the small satellite development project, and that part of the laboratory was left unused.

By mentioning this I certainly don't mean to imply that an impressive and ambitious project like Planet Labs could have started in Terni, but in all honesty, I cannot rule it out. Also because there are garages and small labs all over the world, along with young and curious minds. Planet Labs' founders came from an experience gained in the context of a national institution that has always played a leading role in the space industry: NASA. We too came from the world of universities and research, and had worked directly with NASA and INFN for more than a decade.

In both cases we are talking about a group of motivated researchers that would grow over time, attracting other bright and curious young people. The difference was evident when it came to making the leap to a larger project in terms of planning and investment. As happens all too often, while the American team quickly found their first private investors, the Italians—lacking a broad network of venture capital investors, fell back on institutional financing; however, receiving no meaningful response from that sector, the opportunity was lost. Ten years later I found myself presiding over the ASI and one of the most important programs we decided to launch during my management tenure involved developing a national, industrial supply chain for small, high-tech satellites. The goal, now achieved, was to establish and expand Italy's capacity in this crucial sector. Planet Labs was already a global success, but the research and experimentation field was so promising that there was still room for other, well-structured and well-planned initiatives. I know that in 2015 other European nations envied the timing of this ASI initiative. However, I am still convinced that it would have been much better to have seized the opportunity ten years earlier, at the dawn of this area of space exploration. When dealing with innovative, breakthrough ideas, there is always a time when the prospects of success are not yet clear and the risk of failure seems to be very high; on the other hand, the initiative of those who make the first move has the best chance of being rewarded, as was definitely the case with Planet Labs.

33

THE DAWN OF THE FUTURE

The image that is the common thread running through these pages, dawn, can be associated with the passage from darkness to light, from that which does not exist to that which does. But it also represents the transition from the present to the future and from ignorance to knowledge, which is a bit like what happens every day thanks to research and scientific progress. Starting from the description of the beginning of the cosmos, we have come to the present time. Let us now try to take a further step forward, attempting something very difficult: understanding where we are going and what remains to be discovered.

As I had the opportunity to explain in the previous chapters, it is necessary to start from a fact we already know: the panorama of the universe surrounding us is boundless, in time and in space, and our presence seems quite small. Should we therefore consider ourselves insignificant? I would say not; quite the opposite. The truly extraordinary fact is that we are aware we exist. It is our ability to relate intelligently to reality that can ensure us, as a living species, an enduring future. However, we have to do so in such a way that humanity can reach this future in the right conditions. Unfortunately, in this sense there are no guarantees. Over the last century the human presence on the planet has substantially altered the environmental balance. It's enough to recall that the level of energy produced and consumed by humans is now comparable to that which keeps the Earth's tectonic plates in motion. As has been amply demonstrated, the effects of our existence on the planet are no longer negligible. This is not the first time such a thing has happened. Think about what happened to cyanobacteria—the microorganisms that dominated the ecosystem up to about 3 billion years ago by producing oxygen from carbon dioxide—until the oxygen in the atmosphere became so abundant that it was, substantially, the cause of their extinction.

The truth is, we don't have much time left. If we do not take concrete action to correct our aim in the next few decades, the consequences will be disastrous. Climate change, in particular, will take us back by about 3 million years, to a time when *Homo sapiens* did not yet exist. It will be our ability to develop technologies capable of reducing our environmental impact, and the adoption of behaviors compatible with the Earth's available resources, which will decide our survival. So, we have a chance, but only if our intelligence can control the most regressive and harmful of our instincts and if politics and society are farsighted enough to be guided by science.

Since it first appeared, about 350,000 years ago, our species has survived and slowly developed by waging a continual battle against much stronger and more resilient animals, completely absorbed by the need to meet our primary requirements—such as food and reproduction—and involved in perennial, violent conflicts between different groups, always at the mercy of nature: mysterious, threatening, and unpredictable. There is no doubt that, from an evolutionary point of view, intelligence has made a difference in guaranteeing our rapid diffusion across the planet. But we are used to fighting for ourselves, not against ourselves. Our goal has been to broaden our territory and our sphere of action, not to restrict them. Limits have been imposed on us from the outside, but we have always tried to overcome them, both as individuals and as a society. What is certain is that in our history, however short compared with that of the universe, we have never lived in circumstances of such extraordinary abundance. Even if there are still significant areas and populations that are below the poverty line, today billions of people have access to better education, medical care, food, and housing than ever before. However, all of this requires a much more rational and efficient use of resources compared with what we've done in the past.

Let's assume that we can manage the environmental issues well enough that we make it to the end of the twenty-first century, and perhaps into the twenty-second, without global disasters. If the last century was the century of physics, astrophysics, and cosmology, one wonders what the science of the twenty-first century will be. Perhaps biology, which thanks to DNA sequencing will reveal the mechanisms of life, with a potential for manipulation that could substantially change the destiny of humankind? Or will space be the new continent of the future, with the possible colonization of the Moon and Mars, the robotization of other planets inaccessible to us, and the exploration of nearby stars? Or will it be the century of artificial intelligence (AI), used to massively enhance scientific computing and thus extending our understanding of the world? In 2017, Google's Deep Mind algorithm, AlphaZero, independently learned how to play chess in just four hours and then trounced Stockfish, the world's most powerful CPU chess engine. AlphaZero performs 1,000 times fewer operations per second than Stockfish, but has the advantage—brace yourselves—of not having been programmed by a human. It is completely "free" to learn the best strategies to make its moves and checkmate its opponent, designing its own strategies that we may never know.

In the twenty-first century, all of this will probably happen at the same time. The dawn we see is already leading us toward a world where—barring serious unforeseen events such as global wars or natural disasters—scientific and technological innovation will continue to grow exponentially, but the results will be predominantly managed by exploiting sophisticated computing systems.

In physics, for example, we will see the emergence of laws that apply to complex systems emerging more frequently. We will approach the understanding of and control over nonlinear phenomena, those in which the effect is not proportional to the cause. As for computers, I am sure that they will help us more and more in those areas of science in which we are exploring or simulating a huge number of specific cases or analyzing mountains of data, but I doubt that they will play a decisive role in coming up with new ideas and visions, the only kind we really need to understand the foundations of the physical world. In short, understanding the mysteries of dark energy and dark mass will require the creativity and originality of a new Einstein rather than the power and systematic skills of AlphaZero or, perhaps, will require a collaboration between the two.

The life sciences are also an area where progress will be substantial and constant. As we know, DNA manipulation techniques and our understanding of DNA's role in cellular processes has opened the door to an infinity of developments. Here the role of AI will increase substantially. The number of cases and contexts connected to biology and medicine is enormous. Considering that biological mechanisms are based on a solid substrate of elementary chemical processes, help from computers will be decisive. Furthermore, the incredible power of genetics will soon bring us face to face with a series of fundamental ethical questions, the most important of which is the one that affects our mortal destiny. At the same time, we have to agree on the threshold for manipulating the genetic code of humans and animals. Once again, in this case I'm afraid that computers will do little to clarify our ideas.

As for space and its exploration, we mentioned the rapid evolution of the space industry and the prospects for its exploration by both humans and robots. The technologies needed to create human colonies on other planets, and to undertake the requisite long journeys into deep space, will require substantial improvements in the equipment currently available. It is actually very complicated to intervene if something goes wrong or breaks, and it is impossible to ensure perfect quality in systems that are too complex; we have to invent technologies capable of correcting errors or dealing with unexpected events autonomously. In short, we need technologies able to repair themselves at both the microscopic and the macroscopic levels and in terms of software and hardware—just like the living world does, repairing wounds, fractures, sickness, and radiation damage, for example. This is how *Avalon*, the spaceship in the sci-fi film *Passengers* works (anyone who has seen the film will know exactly what I'm talking about). It is built to transport 5,000 people in a state of cryopreservation and can travel for 150 years in full autonomy, controlled by its computer. *Avalon* was designed to function as a living being and is able to continuously maintain and repair itself. Too bad that on Earth—and in real life—such technologies are almost nonexistent. We are still to used to living in a society oriented toward the mass consumption of disposable products. Only when we have these technologies available will we be able to calmly face a journey to Mars or other destinations in the solar system that offer the promise of water and other resources, such as the icy moons of Jupiter or Saturn.

Some argue that we should explore space to resolve the Earth's sustainability problems, certain that space technologies will allow humanity to find a better life on another planet. Once again, we must not delude ourselves. Do we really think that we can delegate to other systems or circumstances (powerful computer networks, alternative planets, unparalleled technologies) the choice of which direction to follow? Will these other systems take our place in deciding our goals? I believe that only humankind can find the answers to the questions that are intimately linked to our human essence. It's true that the resources present in the cosmos are, in fact, unlimited, but the way to reach them can be mediated only through us—our goals, our priorities. Above all, it will be necessary to provide education and culture to billions of people who, for the most part, still do not have access to it; then accompany them along the path toward social inclusion so that they become not only participants in but also beneficiaries of the exponential technological growth. We would do well to base our coexistence on a new humanism. Future generations must be able not only to free themselves from an economy and politics tied to war and natural disasters but also to successfully manage the power of knowledge, uniting it with rather than separating it from serious reflection about our common destiny.

It might seem strange to focus on the centrality of humankind in the future of knowledge, but it's for a good reason. Thanks to a series of slow transitions and continual transformation, we have evolved through societies made up of hunters, animal breeders, farmers, artisans, workers, clerks, and engineers. Since the industrial revolution, the advent of machines has multiplied the effectiveness of human work and the quantity of goods produced by each worker, in agriculture and industry as well as in the service sector. In recent decades, machines have benefited from the addition of computers, further increasing the flexibility and quality standards of human labor. Artificial tools have outclassed human skills in almost every industry, and are now moving into the most sensitive and central of sectors, that which is governed by human intelligence. As we have seen, machines already surpass us in chess and other games, such as the traditional Chinese game Go, considered an exclusively human domain due to the creativity and strategic intuition required.

Research in this area is even working on understanding the neurophysiological foundations of our intellectual abilities. The European Commission's Human Brain Project, for example, is a billion-euro research program that aims to reconstruct the functioning of the brain starting from the molecular scale. It seems as if this will be difficult to achieve, also because the characteristic complexity of our brains does not seem to be within the reach of even the most powerful computers. The point is, this is also what was said about chess, but we have seen how that turned out in the end.

The real question is another, much more profound one: it concerns the role of our species in the evolution of life on Earth. To what extent are we masters of our own destiny? To what extent do we find ourselves in the intermediate stage of a development that began billions of years ago and that, one day, could relegate us to a secondary role relative to machines? Will our species end up like the lumberjack in the cartoons, the one who severs the very branch being sat upon? Or will we succeed in colonizing new star systems with the help of ever more powerful machines, controlled by us? Here I have no hesitations; I believe that the extraordinary plasticity and adaptability demonstrated by human intelligence justifies an evolutionary path in which the contribution of artificial intelligence will be at our service and not vice versa. In fact, this human–machine synergy will enable us to pursue many other objectives that are unimaginable today. However, it will not be an easy road ahead. Before reaching that condition, we will see an impressive series of radical cultural, behavioral, and economic changes. This is why the new humanism, which we must develop very quickly, requires as much philosophy as it does technology and as much ethics as it does science.

All that said, what, then, is left to discover? It is a subject impossible to treat exhaustively. We can say that some objectives are quite evident; others are, more than anything else, desires or visions.

The solution to the mystery of dark matter falls within the first category. If we are not gravely mistaken, there must be an answer to what about 80 percent of the mass of the universe is. Even the study of the physical vacuum's properties will hold some surprises for us. Once again we are grappling with an enormous amount of energy at the cosmic level, regarding the nature of which we still know very little. Another objective is to find an overarching scheme that includes the extremes, left open, in space and time, of the universe in which we live. It is necessary to proceed toward Planck's microscopic world, dominated by quantum mechanics, to understand if space and time are distinct. Looking at the infinitely large, one day we can hope to see technologies capable of folding the fabric of space-time to access otherwise unreachable areas of the universe. On the other hand, being confined to a remote corner of the universe cannot be the great aspiration of a species of explorers such as our own. We have the strong desire to be able to travel across the universe, exploiting the laws of general relativity like what we saw in the movie *Interstellar*.

Until that day comes, the search for the origin and presence of life in the universe continues to be a central issue in the twenty-first century. Also because, there's no getting around it: there is no known and valid reason to suppose that the Earth is the only place life has developed. Based on what we do know (but we really know too little), life in the universe could also exist in abundance. In all likelihood, in the brief visits that humans have made to the planets and satellites of the solar system, it has, for now, escaped us. These are just examples of everything that might await us in the future, and that will help us to understand how, up to now, we have taken only a few small steps in the direction of knowledge. We are beginning to understand that, in the end, the universe and its infinite spaces are not so strange; perhaps it is we, who consciously observe them, who are.

ACKNOWLEDGMENTS

It would not have been possible to write this book if I had not had the good fortune to meet, over the course of my life, many people who shared their experience and knowledge with me, in a reciprocally contagious curiosity and passion for science. It is impossible to mention them all, but I am deeply grateful to each of them. I would especially like to thank Enrico Alleva, Pietro Battiston, Giovanna Costanzo, Bruno Giacomazzo, Roberto Iuppa, Massimiliano Rinaldi, and, with gratitude squared, Maria Prodi, for her help with corrections, suggestions, and discussions during the process of drafting this book.

INDEX